Self-Assessment: Teaching Questions for MRCP(UK) and MRCP(I) Part II Written Exams

Second Edition

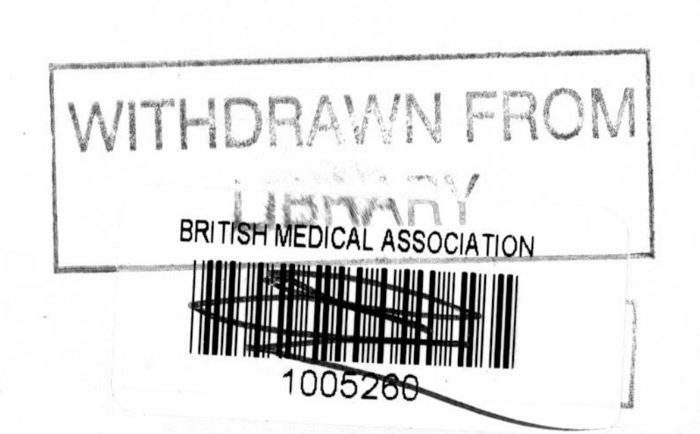

This page was intentionally left blank

Second Edition

Self-Assessment: Teaching Questions for MRCP(UK) and MRCP(I) Part II Exams

Osama S. M. Amin
MD, FRCP(Edin), FRCP(Glasg), FRCPI, FRCP(Lond), FACP, FAHA, FCCP

Clinical Associate Professor of Neurology
International Medical University
Kuala Lumpur, Malaysia

Dr. Osama Shukir Muhammed Amin
Clinical School, International Medical University,
Jalan Rasah, 70300 Seremban, Negeri Sembilan,
Malaysia
Email: dr.osama.amin@gmail.com, osamashukir@imu.edu.my

First Edition: 2009
Second Edition: 2017
ISBN: 978-1-365-77114-9

Distributed by Lulu Press, Inc. Northern Carolina, USA.

Dedication

To my lovely family:

Sarah, Awan, and Naz

Acknowledgements

I woul like to thank my dear patients; their real clinical scenarios were used to formualte and generate these questions.

A special gratitude goes to my other half, wife Sarah, for her endless support and encouragement, and of course her extreme patience.

Osama

Preface

"The best preparation for tomorrow is to do today's work superbly well", Sir William Osler (1849-1919).

This is the 2nd edition of my previous book, "Self-Assessment for MRCP part II; 135 Best of Many Questions with Photographic Materials for MRCP(UK) and MRCP(I) Part II Written Exams", which was published in 2009. The book has undergone several changes and updates. Membership diplomas of the Royal Colleges of Physicians of the United Kingdom, MRCP(UK), and Ireland, MRCP(I), have several things in common. The best initial step is to read accredited textbooks. After then, self-assess yourself. Read this book, chapter after chapter, try to find out your mistakes and gaps in knowledge, and then re-read what you have missed.

This book is a rapid self-assessment tool. There many clues and tricks on how to analyze and answer certain questions. Each chapter does not contain more than 15 questions; it's like a mock paper/exam. I have included many data interpretations and photographic materials. Each question has an *objective*. Try to target that objective.

In writing this book, I tried my best to include the commonest examination themes that have been encountered recently. Undoubtedly, if you are well-prepared, you will pass the examination very easily. No need to panic when you hear your colleagues' past experiences. Lack of preparation is the single most common reason for failure. Remember, practice makes perfect. Read and self-assess, that's it!

For a more comprehensive self-assessment, try "*Neurology: Self-Assessment for MRCP(UK) and MRCP(I)*".

Good luck with your career and exams!

Osama S. M. Amin
February 2017

Recommended Reading

1. Walker B, Colledge NR, Ralston S, Penmen I. *Davidson's Principles and Practice of Medicine, 26th edition.* London: Churchill Livingstone; 2014.

2. Kumar P, Clark ML. *Kumar and Clark's Clinical Medicine, 8th edition.* Philadelphia: Saunders Ltd.; 2012.

3. Longo D, Fauci A, Kasper D, Hauser S, Jameson J, Loscalzo J. *Harrison's principles of Internal Medicine, 18th edition.* New York: McGraw-Hill Professional, 2011.

4. Dale DC, Federman DD, *ACP Medicine, 3rd edition.* Philadelphia: BC Decker Inc.; 2007.

5. Barrett KA, Barman SM, Boitano S, Brooks H. *Ganong's Review of Medical Physiology, 24th edition.* New York: McGraw-Hill Education / Medical; 2012.

6. Lang TA, Secic M (eds.). *How to Report Statistics in Medicine, Annotated Guidelines for Authors, Editors, and Reviewers, 2nd edition.* Philadelphia: The American College of Physicians, 2006.

7. Rosene-Montella K, Keely EJ, Lee RV, Barbour LA (eds.). *Medical Care of the Pregnant Patient, 2nd edition.* Philadelphia: The American College of Physicians; 2007.

Table of Contents:

Chapter 1: Cardiology..11

Chapter 2: Respiratory Medicine..37

Chapter 3: Clinical Pharmacology, Therapeutics, and Toxicology...65

Chapter 4: Pharmacology, Therapeutics, and Toxicology............91

Chapter 5: Hematology, Oncology, and Palliative Medicine.........109

Chapter 6: Rheumatology and Diseases of Bones.....................129

Chapter 7: Endocrinology and Metabolic Medicine..................149

Chapter 8: Infectious Diseases and Genitourinary Medicine.........173

Chapter 9: Gastroenterology and Hepatology....................... 189

Chapter 10: Nephrology..211

Chapter 11: Dermatology...225

This page was left intentionally blank

Chapter 1

Cardiology
15 Questions

This page was left intentionally blank

1) A 65-year-old retired police officer has been experiencing central/substernal chest pain over the past 5 months. The patient says that his chest pain occurs when he walks 2 blocks on the flat level and against hill as well as when doing the stairs. It is relieved by resting for about 10 minutes. The blood pressure has always been high, as his GP told him. He does not take his oral anti-diabetic medication regularly. He smokes 2 packets of cigarettes per day and drinks 3 units of alcohol at weekends only. Past surgical history reveals right-sided total hip replacement 2 years ago. There is no family history of note. His daily medications and dosages are Lisinopril 10 mg and Metformin 1000 mg. Examination reveals a BMI of 28 Kg/m^2, regular pulse rate of 90 beats/minutes, and a blood pressure of 170/100 mmHg. The neck is supple and the abdomen is benign. His chest shows features of COPD. The left knee is swollen and is painful on movement; there is crepitus. 12-lead resting ECG that was done by his GP last week shows marked left ventricular hypertrophy. You have ordered some blood tests to be done.

What would you do next to elucidate the cause of his chest pain?

a) Treadmill exercise ECG testing

b) Coronary angiography

c) Cardiac electrophysiological studies

d) Myocardial perfusion imaging

e) Repeat his 12-lead resting ECG after 1 week

Objective: Review the diagnostic approach of coronary artery disease.

A common theme in the MRCP examinations is the diagnostic approach of ischemic heart disease. This man has many risk factors for coronary artery disease (CAD) and his chest pain sounds ischemic. His resting 12-lead ECG is already abnormal (but does not reflect myocardial ischemia) and repeating the same test now would add nothing.
Can we proceed with formal exercise stress ECG testing? The answer is no; we cannot do formal exercise ECG testing in the following 2 groups:

1. Patients who are unable to exercise sufficiently due to leg claudication, arthritis (as in our patient), deconditioning, or associated pulmonary disease. A diagnostic test that fails to achieve 85 to 90% of the patient's predicted maximal heart rate is considered inadequate to rule out ischemic heart disease if the test is otherwise negative (i.e., absence of chest discomfort or ECG findings).

2. Patients with ECG changes at rest that can interfere with interpretation of the exercise test. These abnormalities include pre-excitation (Wolff-Parkinson-White) syndrome, a paced ventricular rhythm, more than 1 mm of ST depression *at rest*, complete left bundle branch block, and patients taking digoxin or with ECG criteria for left ventricular hypertrophy, even if they have less than 1 mm of baseline ST depression. On the other hand, patients with right bundle branch block or those with less than 1 mm of ST depression at rest are candidates for diagnostic exercise ECG testing in the appropriate clinical setting.

Pharmacologic stress testing with radionuclide myocardial perfusion imaging or echocardiography is convenient in patients who cannot exercise using a treadmill or a bicycle; exercise testing with radionuclide myocardial perfusion imaging or echocardiography should be performed in patients with resting ECG abnormalities, and an imaging study should be performed to localize ischemia or assess myocardial viability.

Exercise ECG testing is most useful in patients with an *intermediate* pretest probability (variably defined as between 25 and 75% or between 10 and 90%). Positive tests in these patients are more likely to be true positives due to the predictive accuracy of the test. The test is less useful in patients with a high (our patient has a high pretest probability) or low pretest probability. Men over the age of 40 years and women over the age of 60 years with a history of typical angina pectoris already have very high pretest probabilities of CAD based upon symptoms alone (87 and 91%, respectively). Exercise testing is not needed for diagnosis in these patients.

On the other hand, a relatively high false positive rate may be expected in patients with pretest probabilities of CAD below 25% (e.g., men under the age of 40 years or women under the age of 50 years with atypical chest pain). **(*Correct Answer: b*).**

2) A 69-year-old retired dietician presents to the Emergency Department with central chest pain for the last 6 hours. The pain is partially responsive to many sublingual nitroglycerin tablets. The patient's sister says that her brother's pain has built-up gradually while he was reading the newspaper and he decided to visit the Emergency Room after 6 hours. He developed 3 short-lived anginal attacks within the past 24 hours. The man has chronic stable angina, hypertension, hypercholesterolemia, and type II diabetes. He smokes about 10 cigarettes per day but does not drink alcohol. The patient's father died of myocardial infarction at the age of 71 years. The man's doctor prescribed atenolol, aspirin, and isosorbide dinitrate for his heart problem, 2 months ago. 12-lead ECG shows deep symmetrical T-wave inversion in the precordial leads from V1 to V6 but there is no Q-wave. Serum troponin-I is minimally raised. The plain X-ray chest film is unremarkable, as are his blood tests. Precordial examination is unhelpful and his blood pressure is 150/95 mm Hg. In the Emergency Room, he has been given Morphine, Aspirin, Clopidogrel, Heparin, Metoprolol, Abciximab, and Nitroglycerin infusion, and he is now stabilized.

Which one of the following is the best you can do for this patient?

a) Intravenous alteplase

b) Modified exercise ECG stress testing

c) Coronary angiography and revascularization

d) Intra-arterial lepirudin

e) Oral Ramipril

Objective: Review the management of acute coronary syndromes.

This patient has developed non-ST segment elevation myocardial infarction (NSTEMI) with a TIMI score of 5 out of 7; this would put him in the high-risk category. NSTEMI patients with intermediate (TIMI score of 3-4) or high (TIMI score of 5-7) risk benefit from early coronary intervention once the patient is stabilized on optimal medical therapy. High-risk patients should also receive Clopidogrel and GPIIb/IIIa inhibitors (such as abciximab). Direct thrombin inhibitors (e.g., Lepirudin) are suitable alternatives to unfractionated heparin when the latter is contraindicated (e.g., history of heparin-induced thrombocytopenia).

Thrombolytic therapy has no place in NSTEMI. ACE inhibitors can be used when indicated (e.g., to control blood pressure, tackle congestive heart failure...etc.) but by their own they are not part of the treatment of NSTEMI. Stress ECG testing can add prognostic significance, but is usually deferred until the patient stabilizes after at least 1 month. (***Correct Answer: c***).

3) A 31-year-old woman visits the physician's office because of poor exercise tolerance. She was diagnosed with NYHA functional class III idiopathic dilated cardiomyopathy 7 months ago, after developing breathlessness, orthopnea, raised JVP, bi-basal crackles, and leg edema. Currently, she receives Enalapril 10 mg twice daily, Carvedilol 12.5 mg twice daily, Spironolactone 25 mg once daily, and Frusemide 40 mg once daily. She is compliant with her regimen and she denies the ingestion of any other medication or doing drugs. On a specific enquiry about her current complaint, she answered, "I feel weak when I walk on the flat after some time, doing the stairs, scrubbing or mopping the floor, or after dusting the furniture." Examination reveals mildly raised JVP, trace leg edema, clear lung bases, and no gallop rhythm. ECG shows wide-spread non-specific ST-T segment changes and QRS complex duration of 170 msec. Her up-to-date ejection fraction is 34%. The blood pressure is 100/60 mmHg, pulse rate is 62 beats/minute, which is regular in rhythm and volume.

What would you do to improve her exercise tolerance?

a) Give Amiodarone

b) Start Digoxin

c) Arrange for resynchronization therapy

d) Refer for cardiac transplantation

e) Increase the dose of Frusemide

Objective: Recognize indications and benefits of resynchronization therapy in congestive heart failure.

Biventricular pacing (resynchronization therapy) has been shown by many studies to improve cardiac pump and contractile functions and can prevent ventricular remodeling; ventricular dys-synchrony exacerbates LV dysfunction. This form of therapy is used in highly selected patient with:

1. Advanced heart failure (NYHA functional class III or IV).
2. Severe LV dysfunction (ejection fraction <35%).
3. Intra-ventricular conduction delay (QRS complex duration >120 msec).

This patient's medical treatment (ACE inhibitor, beta blocker, aldosterone antagonist, and loop diuretic) has resulted in some degree of symptomatic improvement; however, she is a good candidate for resynchronization therapy for a better response to occur.

Digoxin may reduce hospitalization rate and may produce some symptomatic improvement even in patients with sinus rhythm but has no survival benefit; however, it is unlikely that it will produce any substantial improvement in this woman. The question did not address any form of cardiac arrhythmia; Amiodarone use is not justified. Cardiac transplantation is an option in refractory heart failure (which is not the case here). Although her JVP is mildly elevated, her chest has no basal crackles and her legs demonstrated trace leg edema; the volume status is reasonably well-controlled and there is no need to increase the diuretic dose. (***Correct Answer: c***).

4) A 68-year-old man comes in for his annual check-up. He has long-standing hypertension and type II diabetes mellitus. He says that he has difficulty doing the stairs and shopping and has noticed that he has no power when he walks every morning. He lives alone in a 2-story house. He is an ex-smoker but drinks a glass of wine every night. He thinks that his condition is deteriorating in spite of being compliant with his daily 50 mg Atenolol and 100 mg Aspirin and he is desperate for help. The patient's physique is thin. The blood pressure is 170/110 mmHg and the pulse rate is 74 beats/minute. Blood tests show HbA1c 10%, LDL-cholesterol 120 mg/dl, blood urea 40 mg/dl, and Hb 13 g/dl. Echocardiographic studies reveal concentric left ventricular hypertrophy and ejection fraction of 44%. You are awaiting his chest plain X-ray film.

Which one of the following you would *not* choose with respect to this man's management?

a) Add Enalapril

b) Increase the dose of Atenolol

c) Add Pravastatin

d) Do urine examination

e) Prescribe Glyburide

Objective: Review the clinical presentation of systolic dysfunction and its drug therapy.

This man has poorly controlled hypertension and hyperglycemia (note the high HbA1c) and has developed heart failure (ejection fraction of 44%). He should have his blood pressure well-controlled and his cardiac failure well-tackled. Atenolol, a beta blocker, is cardiac depressant and should be stopped gradually while anti-failure measures (ACE inhibitors or ARBs, diuretics, digoxin, spironolactone, and "certain" beta blockers) are being introduced. The question did not address any specific anti-diabetic medication he takes, and you should conclude that his diabetes is being managed by diet alone, which has not achieved a target HbA1c <7%; therefore, glyburide can be introduced in this thin man to lower his hyperglycemia. Diabetes mellitus is a coronary artery disease equivalent and his LDL-cholesterol should be kept below 100 mg/dl by drug therapy (such as statins). Type II diabetic patients should have their urine tested annually for proteins in addition to retinal examination.

Note that Metformin and Thiazolidinediones (TZDs) are contraindicated in heart failure. ***(Correct Answer: b)***.

5) A 22-year-old college student has been referred to your Emergency Department from a general practitioner's office as a case of myocardial infarction. The patient says that he visited the GP because of precordial pricking pain and tenderness. The GP did an ECG for him and found ST-segment elevation in leads V1-3 in addition to right bundle branch block. The patient's two older brothers died suddenly during their 3rd decade. The patient is frightened and is desperate for help. He denies retrosternal chest pain or shortness of breath. There are no palpitations. He does not smoke but drinks beer at weekends. Examination shows blood pressure of 125/78 mmHg and regular heart rate of 94 beats/minutes. The neck is supple, the chest is normal, and the abdomen is benign. 21-lead resting ECG confirms the GP's finding. His blood tests are: Hb 15 g/dl, blood urea 36 mg/dl, serum potassium 4.2 mEq/L, and plasma glucose 92 mg/dl. Serum troponin-I and MB-CK enzyme are normal. Chest X-ray film is unremarkable.

What is the best step for the time being?

a) Thrombolytic therapy

b) Coronary angiogram

c) Aspirin

d) Reassurance and discharge

e) Abciximab

Objective: Review Brugada syndrome and its characteristic ECG findings.

This young man has no risk factors for coronary artery disease. Precordial pricking pain is not cardiac in origin. The presence of "asymptomatic" ST-segment elevation and RBBB in leads V1-2 in someone with strong family history of sudden death is highly suggestive of Brugada syndrome. Reassurance and explanation is all that is required for the time being and the patient should be advised about the placement of a cardioverter-defibrillator. The other stems are used in acute coronary syndromes. The vast majority of cases of sudden cardiac arrest (SCA) and sudden cardiac death (SCD) due to ventricular fibrillation (VF) are associated with structural heart disease, particularly coronary heart disease (CAD). SCA in the normal heart is an *uncommon* occurrence, accounting for 5% of cases only.

Some causes of SCA in patients with apparently normal hearts have been identified; these include Brugada syndrome, long QT syndrome, pre-excitation syndrome, and commotio cordis (Latin, "agitation of the heart").

Brugada syndrome is associated with a peculiar pattern on ECG, consisting of a *pseudo*-RBBB and persistent ST segment elevation in leads V1-3. Three different patterns of ST elevation have been described:

In the classic Brugada type 1 ECG, the elevated ST segment ($\geq$ 2 mm) descends with an upward convexity to an inverted T wave. This is referred to as the "coved type" Brugada pattern. The type 2 and type 3 patterns have a "saddle back" ST-T wave configuration, in which the elevated ST segment descends toward the baseline, then rises again to an upright or biphasic T wave. The ST segment is elevated $\geq$ 1 mm in type 2 and <1 mm in type 3. (***Correct Answer: d***).

6) A 32-year-old man visits the physician's office for a scheduled annual visit. He was diagnosed with hypertrophic cardiomyopathy 2 years ago, after a self-referral because of sudden death of 3 family members. He denies syncope, palpitations, exertional breathlessness, or chest pain. However, he admits to having precordial thumping beats every now and then. 24-hour Holter monitoring fails to demonstrate sustained or non-sustained ventricular tachycardia. Echocardiography shows marked LVH but no significant LV outflow obstruction. Exercise ECG testing using a treadmill was inconclusive but he developed hypotension during submaximal exercise. He takes daily Bisoprolol tablets. The man does not smoke or participate in competitive sports.

What is the best action you should do with respect to this man's illness?

a) Add disopyramide

b) Send for radiofrequency catheter ablation

c) Repeat exercise ECG testing

d) Do coronary angiography

e) Advise for placing a cardioverter-defibrillator

Objective: Review the management of hypertrophic cardiomyopathy and its risk of sudden death.

Patients without symptoms should be evaluated by echocardiography, Holter monitoring, and exercise testing; asymptomatic patients generally have a *good* prognosis. Although this man is asymptomatic, he has marked LVH, hypotension upon exercise (abnormal vascular response), and a strong family history of sudden death; these would put him in the high-risk category of sudden cardiac death. The normal blood pressure response to maximum treadmill exercise testing includes at least a 20 mmHg increase in systolic blood pressure from rest to peak exercise. However, 20 to 40% of patients with hypertrophic cardiomyopathy fail to augment their baseline blood pressure during exercise; in some of these patients, the blood pressure falls below baseline values during or immediately following exercise. The best approach in this man is to continue his oral beta blocker and to place a cardioverter-defibrillator.

Patients should avoid vasodilators, Digoxin, and heavy unaccustomed exercises. Needless to say, regular follow-up should be encouraged. (***Correct Answer: e***).

7) A 12-year-old girl presents with short-lived palpitations associated with presyncope. She has pansystolic murmur at the left lower sternal border that increases with inspiration. The mother says that her daughter's murmur was detected by a pediatrician several years ago. The child's mother has bipolar disorder and her father died at the age of 30 years because of drug over-dose. 12-lead resting ECG shows P-pulmonale and right bundle branch block. Echocardiographic examination failed to show vegetations.

Which one of the following with respect to this girl's illness is *incorrect*?

a) Severe cases may show the wall-to-wall heart sign

b) The tricuspid valve is sail-like

c) Stenosis never occur in this girl's tricuspid valve

d) The anterior leaflet of the tricuspid valve is the largest leaflet and is usually attached to the tricuspid valve annulus

e) The right ventricle proper is small and, in some cases, consists of only right ventricular outflow tract.

Objective: Review Ebstein anomaly.

The occurrence of recurrent palpitations (tachyarrhythmia) in the background of tricuspid regurgitation and ECG evidence of right bundle branch block and right atrial hypertrophy should always prompt you think of Ebstein anomaly. Her mother has bipolar disorder and might well have been prescribed lithium in the past. For unknown reasons, Ebstein anomaly occurs with higher frequency in infants of mothers who take lithium during early pregnancy. The anomaly encompasses variable deformities of the tricuspid valve and the right ventricle resulting in a highly variable clinical outcome. The anterior leaflet of the tricuspid valve is the largest one and is usually attached to the tricuspid valve annulus, while the posterior and septal leaflets are usually vestigial or even absent. When the latter 2 are present, they are usually displaced downward, away from the atrioventricular junction and toward the body of the right ventricle or the apex. The edges of these leaflets may be free or attached to chordae. Grossly, the tricuspid valve looks funnel shaped (sail-like) and is variably incompetent; surprisingly, stenosis sometimes occurs.

Because of the downward displacement of the tricuspid valve, the upper right ventricle becomes "atrialized"; the distal portion of the right ventricle, which is referred to as the right ventricle proper, becomes small and, in some cases, consist of only the right ventricular outflow tract. In severe cases, the plain chest films would reveal massive cardiomegaly ("wall to wall" heart) with diminished pulmonary vascularity; the right atrium is prominent and the left heart border becomes straight (or convex) due to dilated and displaced right ventricular outflow.

The chest radiograph may be somewhat normal-looking in patients with less severe disease, however. Wolff-Parkinson-White pattern may be present, which always displays left bundle branch block pattern with predominant S waves in the right precordium; however, supraventricular tachycardia may be present with or without Wolff-Parkinson-White syndrome. Other forms of atrial tachycardia (such as atrial flutter or atrial fibrillation) may be seen in older patients. The phrasing of the question may misdirect you think of atrial septal defect or infective endocarditis. (***Correct Answer: c***).

8) A 29-year-old woman presents poor exercise tolerance, which has been progressing over the last few years. After you interview and examine the patient properly, you do an invasive test. The heart rate is regular. These are her cardiac catheterization findings:

	Pressure (mm Hg)	Saturation (%)
Right atrium (mean)	15	75
Right ventricle	89/25	75
Pulmonary artery	105/40	76
Left atrium (mean)	24	94
Left ventricle	110/3	94
Aorta	112/71	-

All of the following about this woman's illness are incorrect, *except*:

a) This is usually an aftermath of infective endocarditis

b) She has a mild disease

c) Examination of the basal lungs should be unremarkable

d) Pregnancy is safe

e) Warfarin should be given

Objective: Review cardiac catheterization findings of mitral stenosis.

Start by looking at the saturation values; neither step-up nor step-down is found. This would exclude septal defects or shunts within the heart or great vessels (e.g., ASD, VSD…etc.). Then, skim the left-sided findings. There is trans-mitral diastolic gradient of 21 mmHg. This is severe mitral stenosis. Continue and analyze the right-sided pressures; all of them are elevated. This is severe secondary pulmonary hypertension; note that the systolic pulmonary artery pressure approaches that of the aorta! At last, you've made it; severe "rheumatic" mitral stenosis with severe secondary pulmonary hypertension. About 20% of the victims remain in sinus rhythm.

However, anticoagulation is required in severe mitral stenosis (as in our patient), as well as in enlarged left atrium and spontaneous echo contrast or atrial fibrillation. *Severe* pulmonary hypertension per se portends 50% maternal mortality in pregnancy; pregnancy should be avoided. With prominently (and gradually) elevated left atrial pressure, pulmonary edema develops and bi-basal crackles are the expected finding on auscultating the basal areas of the lungs. Mitral stenosis is almost always rheumatic; congenital stenosis, SLE, and carcinoid syndrome are very rare culprits.

The widespread availability of echocardiography has limited the role of cardiac catheterization in the diagnosis of mitral stenosis. However, invasive pressure measurements are warranted if noninvasive tests are not conclusive, there is discrepancy between noninvasive tests and clinical findings or between clinical symptoms and hemodynamics at rest, or severe pulmonary arterial hypertension is out of proportion to the severity of mitral stenosis as determined by noninvasive tests ***(Correct Answer: e)***.

9) A 34-year-old woman presents with poor exercise tolerance over several months. She does not smoke and she denies doing drugs. She takes no medications. Auscultation of the chest is unremarkable. You do cardiac catheterization and this is what you get:

Left ventricular end-diastolic pressure	9 mmHg
Pulmonary capillary wedge pressure	11 mmHg
Pulmonary artery systolic pressure	94 mmHg

What do you think this woman has?

a) Mitral regurgitation

b) Left ventricular dysfunction

c) Severe pulmonic stenosis

d) Thromboembolic pulmonary hypertension

e) ASD primum

Objective: Differentiate between various causes of pulmonary hypertension.

The normal left-sided pressures (including PCWP) would suggest the absence of any hemodynamically-significant lesions there; e.g., mitral stenosis or regurgitation, LV failure…etc. The normal chest auscultation would exclude severe pulmonic stenosis. ASD primum presents at an earlier stage with mitral and/or tricuspid regurgitation. When you find isolated pulmonary hypertension *and* normal chest and left-sided cardiac examination, think of primary pulmonary hypertension or thromboembolic pulmonary hypertension. ***(Correct Answer: d).***

10) A 43-year-old man, who has end-stage renal disease, visits the Emergency Room. He says that he has noticed retrosternal pain that increases with inspiration and swallowing over the past 4 days. ECG shows wide-spread ST-segment elevation. He is given symptomatic treatment and discharged home. One week later, he returns to the Emergency Room short of breath and hypotensive.

What is the cause of the second visit?

a) Extensive myocardial infarction

b) Constrictive pericarditis

c) Acute myocarditis

d) Ruptured papillary muscles

e) Cardiac tamponade

Objective: Review the natural history of acute pericarditis.

This man presented initially with acute uremic pericarditis; note the character of the pain and the ECG findings. Uremic patients are at risk of developing hemorrhagic pericarditis and cardiac tamponade; the latter explains the second visit. ECG alone cannot establish the diagnosis of tamponade, but two findings should raise concern about this possibility:

1. Low voltage with sinus tachycardia: this combination can also occur with chronic obstructive pulmonary disease, pleural effusion, cardiomyopathy, and after open heart surgery.

2. Electrical alternans with sinus tachycardia: this combination can also be seen in cardiomyopathy with severe heart failure.

The physical examination in subacute tamponade may reveal hypotension with narrow pulse pressure, reflecting the limited stroke volume. However, patients with preexisting hypertension may remain hypertensive due to the increased sympathetic activity that is usually present in tamponade. ***(Correct Answer: e).***

11) A 61-year-old man visits the physician's office for a scheduled follow-up. He has chronic stable angina and hypertension. His daily medications are Lisinopril 10 mg and Atorvastatin 20 mg. When he walks 2 blocks on the flat level or does 2 flights of stairs, he develops moderate chest tightness that is responsive to sublingual nitroglycerin. Blood pressure is 110/75 mmHg, hemoglobin 13 g/dl, blood urea 32 mg/dl, serum potassium 4.7 mEq/L, LDL-cholesterol 138 mg/dl, and total cholesterol 210 mg/dl.

What would you do next?

a) Reduce the dose of Lisinopril

b) Double the Atorvastatin dose

c) Stop Atorvastatin and start Simvastatin.

d) Add Ezetimibe

e) Prescribe Isosorbide Dinitrate.

Objective: review the management of hypercholesterolemia in coronary artery disease.

This patient has an established cardiovascular disease and his lipid profile is still above the desired target. Many recent studies have shown that in patients with atherogenic lipid profile, doubling the dose of statin rarely achieves adequate lowering of serum LDL cholesterol (which usually results in a further reduction of 6-10% only), and consideration should be given to adding another agent. In this patient, adding Ezetimibe, 10 mg/day, is a reasonable step to lower the patient's serum LDL-cholesterol to a level below 100 mg/dl (preferably, below 70 mg/dl). ***(Correct Answer: d)***.

12) You have been asked to take care of a 59-year-old man who has just been diagnosed with high-risk unstable angina. Past medical history shows hypertension, hyperlipidemia, left knee osteoarthritis, and gastro-esophageal reflux disease. You admit him to the coronary care unit and do percutaneous coronary intervention with angioplasty. You place 2 stents in the proximal left anterior descending artery. He is now stable and is pain-free. Ejection fraction is 57%. There are mild diastolic dysfunction and concentric LVH. You discuss with him the long-term management plan and his medications list. You prescribe Bisoprolol, Simvastatin, Aspirin, Clopidogrel, Omeprazole, and Captopril. However, one of your colleagues suggests switching one of those medications to something else.

Which one of the following does he suggest?

a) Switching Bisoprolol to Metoprolol

b) Switching Simvastatin to Fenofibrate

c) Switching Clopidogrel to slow-release Dipyridamole

d) Switching omeprazole to Ranitidine

e) Switching Captopril to Candesartan

Objective: Know the potential interaction of proton pump inhibitors and their possible antagonism with Clopidogrel.

This patient has been prescribed many medications and some of them will definitely aggravate his gastro-esophageal reflux symptoms. The caring physician has anticipated this and has given the patient a proton pump inhibitor (PPI), and that is Omeprazole. However, many recent studies have conclusively demonstrated that PPIs antagonize the anti-platelets action of Clopidogrel. Clopidogrel bisulphate is a pro-drug that should be metabolized in the liver's cytochrome *P4502C19* system to its active form. PPIs also share this pathway in the liver. Lowering the dose of the PPI or using Pantoprazole (which may have the least antagonizing effect) instead of other PPIs has *not* been shown to be an effective policy in reducing this antagonism. The best approach in this patient is to switch Omeprazole (i.e., the PPI) to an H2-blocker agent (such as Ranitidine). There is no need to change the other medications. (***Correct Answer: d***).

13) A 50-year-old man visits the doctor's office. He is reasonably well and healthy, as he says, and leads an independent life. He denies symptoms. He exercises regularly, and neither smokes cigarettes nor drinks alcohol. His family history is unremarkable. He is interested in coronary artery disease risk factor reduction and has read in the internet that certain dietary supplements may achieve this target.

You reply positively and advise him to take which one of the following?

a) Folic acid, 5 mg per week

b) Vitamin E, 200 IU per day

c) Vitamin K, 1 mg bi-weekly

d) Fish oil, 1000 mg per day of eicosapentaenoic acid and docosahexaenoic acid

e) Olive oil, 5 ml every morning

Objective: Know the cardiovascular effects of fish oil.

Fish oil, particularly eicosapentaenoic acid and docosahexaenoic acid, has been shown by many studies to reduce the cardiovascular death. The preparation is well-tolerated, as well. Vitamin E and β-carotene preparations actually *increase* the cardiovascular events. Folic acid efficiently reduces the blood's homocysteine levels and demonstrates reduction in the overall venous systemic events; however, there appears no impact on the arterial side of vascular events. (***Correct Answer: d***).

14) Because of poor exercise tolerance, a 62-year-old man visits your office. You ultimately diagnose ischemic cardiomyopathy and give him Ramipril, Metoprolol, Fluvastatin, and spironolactone. You schedule a visit after 12 weeks and he demonstrates favorable response in terms of symptoms. ECG shows QRS duration of 0.09 sec, while the ejection fraction is 49%. LDL-cholesterol is 98 mg/dl.

Which one of the following interventions prolongs this man's life?

a) Add digoxin

b) Prescribe fish oil

c) Arrange for re-synchronization therapy

d) Increase the dose of statin

e) Put a cardioverter-defibrillator device

Objective: Fish oil prolongs life of congestive heart failure patients.

Around 10% of congestive heart failure patients die each year in spite of receiving optimal medical therapy. Omega-3 fatty acids (fish oil), 1000 mg a day, result in enrichment of membrane phospholipids; the latter produces an increment in the arrhythmic threshold, reduction in systemic blood pressure, improvement in arterial endothelial function, and prevention of platelets aggregation. It also favorably affects the autonomic tone. The constellation of these effects result in reduction in cardiovascular death and reduction in hospital admission for cardiovascular events. The QRS complex duration should be prolonged to qualify patients for receiving re-synchronization therapy. The patient does not demonstrate sustained or non-sustained ventricular dysrhythmia and his ejection fraction is above 35%; therefore, the use of an implantable cardioverter-defibrillator device is not justified. Digoxin does not affect the mortality rates of heart failure patients; however, it does reduce the hospitalization rates. Although statins in this patient are part of his secondary prophylaxis regimen against acute ischemic events, these medications do not influence mortality figures in heart failure. (***Correct Answer: b***)

15) Your surgical department colleague consults you about this 62-year-old woman who is planned to undergo laparoscopic cholecystectomy 8 days from today. She has mechanical bi-leaflet aortic valve and takes daily Warfarin. Her INR has always been in the range of 2-3 over the past year. Blood pressure is 135/87 mmHg and pulse rate is 80 beats per minute, which is regular in rate and rhythm. Her up-to-date ejection fraction is 58%. The surgeon requests proper preparation before elective surgery and demands an INR below 1.5.

You tell the surgeon:

a) Stop Warfarin 72 hours before the operation and re-start it again within 24 hours post-operatively

b) It is unwise to discontinue Warfarin; do the operation while the INR is between 2-3

c) Discontinue warfarin 6 days before the operation; admit to the hospital for bridging heparinization 72 hours before the operation; stop heparin 24 hours post-operatively

d) Stop Warfarin 1 day before the operation and start Enoxaparin 12 hours before the operation

e) Stop Warfarin the morning of the operation and restart it again after 5 days

Objective: Review the peri-operative management of anti-thrombotic therapy in valvular heart disease patients.

A very common and problematic situation is the preparation of at risk-of-thrombosis patients (who take anticoagulants) for non-cardiac surgery. An INR of less than 1.5 is safe for most surgical operations in terms of secured hemostasis. The constellation of bi-leaflet aortic valve, absence of LV dysfunction, and presence of sinus rhythm all put this patient at low risk of developing valvular prosthesis thrombosis if the INR is sub-therapeutic. For such patients, most guidelines recommend stopping Warfarin 48-72 hours before the elective surgery (allowing sufficient time for the INR to fall below 1.5) and to resume it within 24 hours post-operatively; there is no need for bridging heparinization (as the risk of thrombosis is low).

The presence of anyone of the following increases the risk of developing mechanical aortic valve prosthesis thrombosis: old generation valves (e.g., Starr-Edward); the presence of another prosthesis (e.g., in the mitral area); LV dysfunction (especially if the ejection fraction is <30%); atrial fibrillation; and prior thromboembolic events. (***Correct Answer: a***).

Chapter 2

Respiratory Medicine
15 Questions

This page was left intentionally blank

1) A 61-year-old journalist visits the physician's office with exertional breathlessness for few months. He has cough with scanty whitish sputum and occasional wheezes. He is life-long heavy smoker. He denies any form of chest pain. Examination reveals hyper-inflated chest, hyper-resonance percussion note, and diminished breath sounds all over the chest. There are few scattered wheezes. The liver edge is palpable but there is no splenomegaly. His bloods show hemoglobin of 9 g/dl, serum potassium 3.0 mEq/L, and leuko-erythroblastic blood picture. He takes many forms of daily inhalers in addition to oral prednisolone but with little relief of his chest problem. Chest X-ray film is consistent with emphysema but there is an irregular rounded opacity, 3 x 2.5 cm, a little above and lateral to the right hilar region. Up-to-date FEV_1 is 50% of the predictive value. Abdominal ultrasound is negative and brain CT scan is normal. Biopsy of the mass is consistent with small-cell lung cancer.

How would you treat?

a) Surgical removal of the mass

b) Endoscopic ablation of the mass

c) Hospice care

d) Chemotherapy

e) Radiotherapy

Objective: Differentiate between the management of small cell and non-small cell lung cancer.

Divide lung cancer into 2 groups: small-cell and non-small cell lung cancers. Small-cell lung cancer is staged into 2 stages according to the Veterans' Affairs Lung Study Group staging system; limited and diffuse (extended). The limited cancer is "limited" to one hemithorax and is "within" the field (port) of radiotherapy (this roughly corresponds to stages I through IIIB of non-small cell lung cancer). This form can be treated with radiotherapy alone, and chemotherapy is an additional *option* in many patients. The diffuse (extensive) variety has diffused "outside" one hemithorax and is "not within" the radio-therapeutic field (port). This stage is best targeted by chemotherapy. Our patient has anemia with leuko-erythroblastic blood picture that might well represent bone marrow infiltration by the malignant cells; this is diffuse-small cell lung cancer and stem "d" is the treatment of choice.

For diffuse small-lung cancer, two-drug combination of etoposide *plus* either carboplatin or cisplatin is used; regimens substituting irinotecan, topotecan, or epirubicin for etoposide are reasonable alternatives. The low serum potassium of our patient might well represent SIADH; a common association, which is usually asymptomatic and is a biochemical diagnosis. Small-cell lung cancer has no surgical treatment with the intent to "cure".

Non-small cell lung cancer encompasses squamous cell carcinoma, adenocarcinoma, and large-cell cancer. This group can be staged (stage I-IV, or TNM) and each stage has it own treatment modality. Surgical treatment is used in localized and locally advanced tumors. ***(Correct Answer d).***

2) A 32-year-old taxi driver is brought to the Emergency Department by his wife. The patient is short of breath. He has had chronic persistent asthma for the last 9 years and is compliant with his daily glucocorticoid and beta agonist inhalers. His wife says that he got flu and his GP prescribed paracetamol and ampicillin for him 3 days ago, but today his breathlessness has increased. He is afebrile and conscious. Pulse is regular at a rate of 100 beats/minutes. Blood pressure 110/60 mmHg. His chest is wheezy but there are no crackles. Chest X-ray film shows generalized hyper-lucent shadow with no pneumonic patch or pneumothorax. He is being given high flow, high concentration oxygen. Blood gas analysis on air is as follows:

PaO_2 8.1 kPa, $PaCO_2$ 4.0 kPa, pH not done.

What is the best next step?

a) High dose intravenous hydrocortisone

b) Montelukast

c) Intravenous magnesium

d) Mechanical ventilation

e) Chest tube insertion underwater seal apparatus

Objective: Review the management of severe asthmatic attacks.

This man has developed a sudden deterioration in his asthma control after an apparently mild upper respiratory tract infection. The question did not mention the PEFR recordings and did not include any clinical life-threatening parameter. However, his $PaCO_2$ is within its normal reference range; CO_2 washout is an expected finding in acute asthma with resulting hypocarbia. This indicates that the patient has impending respiratory failure and going to mechanical ventilation is the correct action to save the patient's life. Montelukast is used in step 4 of asthma treatment, in the long-term, and has no place in treating severe acute attacks. Patients who are still prominently symptomatic while being managed for their acute asthma can receive intravenous magnesium. Oral (or intravenous) glucocorticoids are used in acute asthma attacks but they should not be used alone in the treatment of this patient's condition as a life-saving measure.

The development of unilateral or bilateral pneumothorax is a well-recognized acute asthma complication, which may render the patient virtually unresponsive to conventional treatment. It should always be suspected in acute asthma attacks when patients develop lateralized chest pain (unilateral or bilateral). This patient's X-ray has excluded this possibility. ***(Correct Answer: d)***.

3) A 26-year-old male presents to the Emergency Room breathless. He was watching a TV talk show at home and suddenly developed precordial chest pain with mild shortness of breath. The man ascribed this to his exhaustive daily job as a clerk in a large stock market. Afterwards, the pain increased and the breathless became worse. He smokes 1 packet of cigarettes per day and drinks few glasses of burgundy every few days at night. He denies doing drugs and there are no risk factors for HIV. Family history is unremarkable. Examination reveals obvious distress and breathlessness, blood pressure 80/40 mmHg, pulse rate 130/minutes regular, and hyper-resonance percussion note on the left hemithorax. The chest X-ray film cannot be interpreted due to patient's uncooperativeness. Chest tube is inserted under water seal apparatus on the left side and little air goes out. However, he is still unwell and there is little improvement.

What do you think the next step is?

a) Insert another chest tube on the right side

b) Chest aspiration of the left pleural space

c) Apply suction on the inserted chest tube

d) Give high flow, high concentration oxygen

e) Do 12-lead ECG

Objective: Review the management of spontaneous pneumothorax.

The patient has developed primary spontaneous (non-traumatic) left-sided pneumothorax that gradually has become under tension. Pneumothorax with cardio-respiratory comprise is considered to be under tension regardless of its size; note the patient's hypotension and severe dyspnea. Chest tube should be inserted, and when no improvement is seen (no resolution of the pneumothorax), applying suction on the tube is recommended. There are no cardiac abnormalities and stem "e" can be crossed out. The pneumothorax is in the left pleural space; there is no indicator of bilateral occurrence in order to insert another chest tube within the right pleural space. The pneumothorax is under tension and the patient is unstable and needs immediate relief by chest tube insertion; there is no place for chest aspiration in this patient. ***(Correct Answer: c).***

4) A 57-year-old writer is brought to the Emergency Room with dyspnea. The patient was reasonably well previously. He developed progressive shortness of breath over the past 3 days. He has moderate COPD and receives beta agonist and steroid inhalers. The patient is breathless and uses accessory muscles of respiration. There is a tinge of cyanosis. He is unable to give a proper history. Blood pressure is 110/60 mmHg. There are rapid irregular heart rate and mild pitting leg edema. Blood gas analysis is being done. He has undergone chest X-ray, which is reported to be unremarkable. 12-lead ECG reveals atrial fibrillation (50 to 88 beats per minutes).

Which one of the following statements is *correct* with respect to this man's condition?

a) Pulmonary thromboembolism cannot be excluded

b) Chest tube should be inserted for tension pneumothorax

c) Legionnaires' disease is the likely diagnosis

d) His lobar pneumonia is resolving

e) DC cardioversion should be done

Objective: Review the management of acute COPD exacerbations.

This relatively short scenario is about a middle-aged COPD man who has developed acute exacerbation of his chest problem. But, what is the cause? The unremarkable chest X-ray film has excluded lobar pneumonia, pneumothorax, and pulmonary edema. Mild degree of pitting leg edema may represent cor pulmonale or immobility; however, deep venous thrombosis is still a possibility, which may result in pulmonary thromboembolism. The latter cannot be excluded by the unremarkable chest X-ray film. COPD patients may develop atrial tachyarrhythmia and the scenario did not tell us the duration of the patient's atrial fibrillation. In addition, there is no hemodynamic collapse; therefore, DC cardioversion is not the right choice. Nothing in the scenario points towards Legionnaires' disease (diarrhea, hyponatremia, splenomegaly…etc.). Don't forget myocardial ischemia/infarction as a cause of deterioration in COPD patients. Anyway, lower respiratory tract viral infection is the likely culprit in this man. ***(Correct Answer: a).***

5) A 16-year-old girl visits the Emergency Department because of 3-day history of fever and right upper abdominal pain. You examine the patient and you decide to do chest X-ray. This is her chest plain film.

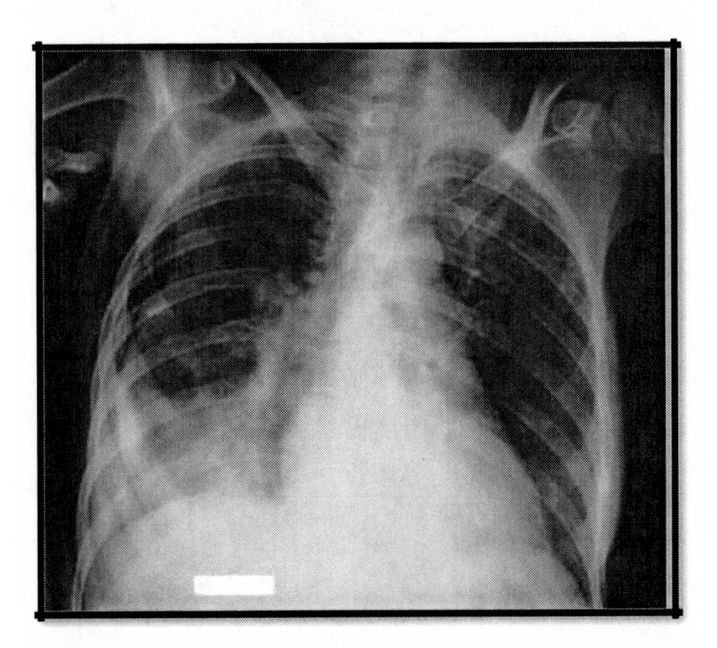

What is the cause of her abdominal pain?

a) Right sub-phrenic abscess

b) Perforated duodenal ulcer

c) Hepatic abscess

d) Right-sided empyema thoracis

e) Right basal pneumonia

Objective: Review the "mode" of presentation of acute pneumonias.

Many basal lung pathologies can masquerade as abdominal symptoms, and some abdominal diseases may come into attention after causing chest problems.

This girl has right basal pneumonic illness. Involvement of the peripheral diaphragmatic pleura would result in pain in the lower right hemithorax and upper right abdomen; sometimes the pain is entirely located in the right upper abdomen. Perforated duodenal ulcer may produce air under the diaphragm; the pain is felt in the epigastrium, which may radiate to the back. Empyema thoracis appears as fluid in the pleural space; obliteration of the right costophrenic angle should be seen in this patient. Liver or sub-phrenic abscess may result in reactionary pleural effusion; these abscesses may appear on the X-ray film as rounded lucent areas with gas inside them. Ultrasound examination or CT scan are the preferred imaging modalities. ***(Correct Answer: e)***.

6) A 34-year-old has been referred to the respiratory outpatients' clinic for further evaluation of progressive shortness of breath. He undergoes pulmonary function testing and the result comes in as:

	Patient's result	Predicted value	% of Predicted
FEV1 (L)	2.4	4.18	57
FVC (L)	3.2	4.80	66
FEV1/FVC %	75	78	
DLCO (mmol/min/kPa)	7.01	10.27	68
KCO (mmol/min/kPa/L)	1.9	1.74	109

All of the following might be responsible for this man's presentation, *except*:

a) Ankylosing spondylitis

b) Myasthenia gravis

c) Bilateral diffuse pleural thickening

d) Idiopathic pulmonary fibrosis

e) Advanced motor neuron disease

Objective: Review the usefulness of pulmonary function testing in various diseases.

You have been provided with many parameters. Scrutinize them and you will find reduced FEV_1 and FVC but normal FEV_1/FVC; this would suggest a restrictive pulmonary defect. You can solidify this by noting the reduced DLCO and you may choose idiopathic pulmonary fibrosis as the cause of this man's dyspnea. But wait! When the reduced DLCO is corrected for the lung volume, the KCO appears within its normal reference range (109 mmol/min/kPa/L in this patient).

Finally, you have reached the correct conclusion; this man has an *extra-parenchymal* pulmonary restrictive defect, not an interstitial one. The potential causes in this man are thoracic cage rigidity (ankylosing spondylitis), poor respiratory muscle function (myasthenia gravis, poliomyelitis, amyotrophic lateral sclerosis), and bilateral diffuse pleural thickening (e.g., following bilateral traumatic hemothoraces or tuberculosis). ***(Correct Answer: d).***

7) A 32-year-old farmer presents with low-grade fever, headache, shortness of breath, dry cough and malaise, around 8 hours after returning from his farm field. He has no such complaints if he takes the day off and is totally asymptomatic during Saturdays and Fridays. He attends the Emergency Room because of dyspnea. His GP told him that this is extrinsic allergic alveolitis. These are his investigations:

Hemoglobin	12.9 g/dl
White cells	9.2 x 10^9/L
Neutrophils	51%
Lymphocytes	20%
Basophils	0%
Monocytes	2%

PaO_2	8.1 kPa
$PaCO_2$	3.0 kPa
DLCO	60% (of the predicted value)
FEV_1/FVC	78%

Blood urea 38 mg/dl

Urine amorphous urate crystals

All of the following about this man's illness are incorrect, *except*:

a) He has an obstructive pulmonary defect

b) His preliminary diagnosis is questionable

c) DLCO is normal if corrected for the patient's age

d) His recent complaints most likely represent viral infection

e) Searching for drug abuse is worthy

Objective: Review the mode of presentation of pulmonary eosinophilic diseases.

If you skim the clinical scenario, you would definitely think of extrinsic allergic alveolitis. The pulmonary function testing reflects a restrictive defect, low DLCO, and type I respiratory failure. The blood urea and urine are normal. So far, the GP's diagnosis is correct. Start analyzing the blood counts. The neutrophils + lymphocytes + monocytes + basophils = 73% of the total peripheral WBCs. This leaves 27% eosinophils! This figure was no shown in the question's data. Striking eosinophilia should cast a strong doubt on the diagnosis of extrinsic allergic alveolitis (hypersensitivity pneumonitis). This question highlights the importance of not relying on others' diagnosis. The physician should analyze patients' investigations vigilantly. ***(Correct Answer: b).***

8) A 43-year-old man visits the doctor's office saying that his asthma has been deteriorating lately in spite of full compliance with anti-asthma medications. This asthma was diagnosed when he was 20 years of age. He admits to having copious amount of offensive greenish sputum every morning. He has bilateral nasal polypi. You find bilateral coarse crackles as well as rhonchi. Chest X-rays shows hyperinflation with fleeting opacities. There is 16% peripheral blood eosinophilia.

What is the reason for this new presentation?

a) Post-primary pulmonary tuberculosis

b) Lung abscess

c) Allergic bronchopulmonary aspergillosis

d) Churg-Strauss vasculitis

e) Cystic fibrosis

Objective: Review the reasons behind progressive worsening of long-term asthma.

The recent production of large amount of offensive greenish sputum in someone with longstanding asthma that has been deteriorating lately should always prompt a search for allergic bronchopulmonary aspergillosis (and proximal bronchiectasis). Nasal polypi are common in asthmatic patients, especially aspirin-sensitive ones. These, together with high eosinophilia and chest X-ray findings, may misdirect you to think of Churg-Strauss vasculitis. Chest X-ray is abnormal in stems "a" and "b". The patient is too old for stem "e", although adult-onset cases have been reported. ***(Correct Answer: c).***

9) A 32-year-old woman has been referred to you as a difficult-to-manage case of bronchial asthma. She was diagnosed with this disease 1 year ago. She has been given salbutamol and beclomethasone inhalators and oral theophylline since then. She says that in spite of using her medicines, she still has poor exercise tolerance and she cannot walk 1 block on the flat without feeling short of breath. She awakes 1-2 times each night dyspneic. She works as a salesperson at a local mall. She lives with her boyfriend and takes oral contraceptives. She denies doing drugs. There is no family history of note. Examination shows a well-kempt woman with a BMI of 27 Kg/m^2 and normal precordial examination. You hear bilateral rhonchi and bi-basal crackles. She has no clubbing or lymph gland enlargement.

How would respond?

a) Do FEV_1/FVC

b) Do peek expiratory flow rate

c) Review the diagnosis

d) Post-bronchodilator FEV_1

e) Trial of oral prednisolone

Objective: Know the features that stand against the diagnosis of asthma.

Superficially, this woman seems to have chronic persistent asthma that is not responding to her current anti-asthma medications. However, the presence of bi-basal crackles should cast a strong doubt on the diagnosis. You should re-evaluate the whole diagnosis. For example, LV dysfunction can simply present as such. ***(Correct Answer: c).***

10) A 61-year-old man has been referred to your office for further evaluation and treatment of cervical spondylosis and severe left hand pain. The patient says that his left arm pain is lancinating and is not responsive to Carbamazepine and Gabapentin. You examine the patient and do some tests. This is one of your investigations.

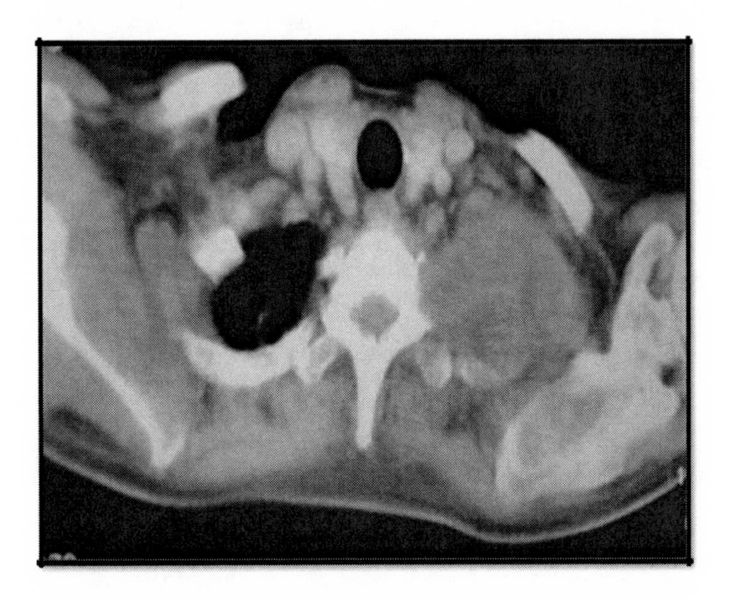

What is the *correct* diagnosis?

a) Retrosternal goiter

b) Fracture dislocation of C7 vertebra

c) Tuberculosis of the spine

d) Pancoast tumor

e) Intramedullary glioma of the cord

Objective: Review Pancoast tumor.

This chest CT scan (at the level of the upper thorax) reveals a soft tissue density mass filling in the left upper lung with some destruction of the transverse process of D1 vertebra.

This picture, together with the refractory neuropathic type of pain this patient has, is highly suggestive of Pancoast (superior sulcus) tumor. The scenario did not mention that the patient smokes (current or ex) and no neurological signs were given. Lesions in the superior sulcus may result in shoulder and arm pain (in the distribution of the C8, T1, and T2 dermatomes), Horner's syndrome, and weakness and atrophy of the muscles of the hand, a constellation of symptoms referred to as Pancoast's syndrome. The majority of patients with superior sulcus tumors present with one or more of these complaints. Due to the peripheral location of the tumor, pulmonary symptoms such as cough, hemoptysis, and dyspnea are uncommon until late in the disease. The overwhelming majority of superior sulcus tumors are non-small cell lung cancers (NSCLCs), mainly squamous cell carcinomas; small-cell carcinomas account for approximately 5% of cases. Although NSCLC predominates as the cause of superior sulcus tumors, less than 5% of all NSCLCs arise in this region. ***(Correct Answer: d)***.

11) A 43-year-old high school teacher visits the Acute and Emergency (A&E) Department. He has been experiencing fever, cough, and chest pain over the past 3 days. He has type II diabetes and hypertension. His daily medications are Glibenclamide and Telmisartan. He smokes around 5 cigarettes per day but does not drink alcohol. You find a temperature of 38.8 °C, respiratory rate of 24, pulse rate of 96 beats/minute, and a blood pressure of 134/81 mmHg. You hear crackles at the lower part of the right chest, posteriorly. The patient is irritable and anxious but is oriented to time, place, and person. Chest X-ray reveals infiltrates in the right lower lung zone. His blood test are:

WBC	13,400 cells/mm^3 (86% neutrophils)
Blood urea	38 mg/dl
Serum creatinine	1.1 mg/dl
Serum potassium	4.3 mEq/L
Serum sodium	140 mEq/L
Random blood glucose	160 mg/dl

Which one of the following is the *correct* statement about this man's illness?

a) Admit him to the intensive care unit

b) Arrange for mechanical ventilation

c) Treat him in the A&E with intravenous antibiotics and then send him home

d) Admit him to your medical ward and give intravenous antibiotics for 3 days

e) Prescribe oral antibiotics and antipyretics and send him home

Objective: Review the indications for inpatient versus outpatient management of community-acquired pneumonia.

This patient has developed community-acquired pneumonia (CAP).
The CURB-65 scoring system helps physicians predict mortality in CAP, determine the safety of outpatient management plan, and the need for hospitalization (and intensive care unit admission).

This scoring system is very easy to use. One point is assigned for the presence of each of the following poor prognostic factors: Confusion, blood URea nitrogen >20 mg/dl (>7 mmol/L), Respiratory rate >30, systolic Blood pressure <90 mm Hg or diastolic blood pressure <60 mm Hg, and age >65 years. Patients who score ≥2 should be hospitalized for initial treatment, and admission to intensive care units is required for those who have a score of ≥3. Our patient scored 0, and he does not display other medical conditions that require hospitalization (such as heart failure, severe hyperglycemia, severe COPD…etc.). He can be managed home with oral antibiotics.

A simplified version (so-called CRB-65), which does not require testing for blood urea nitrogen, may be appropriate for decision-making in primary care practitioner's offices. In this version, admission to the hospital is recommended if *one* or more points are present. Inpatient care for CAP is up to 25 times as costly as outpatient, and many physicians still make inconsistent decisions regarding the need for hospitalization in patients with CAP; avoiding unnecessary hospital admissions helps reduce nosocomial complications as well as reducing the costs. ***(Correct Answer: e).***

12) A 67-year-old retired police officer presents to the Emergency Department with high fever, shortness of breath, productive cough, and pleuritic chest pain. WBC is 16,200/mm^3. Chest X-ray shows a pneumonic patch within the left lower lung zone. You admit him to the hospital and treat as community-acquired pneumonia, using intravenous Levofloxacin in addition to other measures.
On day 4, the patient feels well and takes his daily hospital meals orally. Temperature is 37.4 °C. The patient's dyspnea and cough are improving remarkably. Hemoglobin 14 g/dl, WBC 10,000/mm^3, blood urea 34 mg/dl, and pulse oximetry 94%. You repeat the chest X-ray, which shows the same findings, with no resolution of the pneumonic patch.

What would you do next?

a) Send sputum for acid-fast bacilli

b) Add intravenous Vancomycin

c) Do contrast chest CT scan

d) Continue the same treatment plan for another 3 days

e) Switch intravenous levofloxacin to oral Levofloxacin

Objective: Recall the criteria for switching intravenous antibiotics to oral preparations of the same class in patients with community-acquired pneumonia.

Criteria for switching intravenous to oral antibiotics are: improvement in fever, cough, and dyspnea; and a decrease in the leukocyte count. In addition, the patient must have a functioning gastrointestinal tract and be able to take the oral medication. Our patient has fulfilled all of these criteria and intravenous levofloxacin can therefore be changed to its oral form. In most patients, oral therapy can usually be instituted within 3 days. In general, when transitioning from intravenous therapy, it is recommended that the *same* agent or class of antibiotic be continued by mouth. Discharge from the hospital may be considered when the patient is stable on the oral antibiotic therapy, provided there have been no signs of instability or other medical issues that require in-hospital care.

Chest radiographic findings in acute pneumonias improve in 50% of cases only by the end of the 1st week of optimal therapy. About 70% of patients demonstrate chest radiographic clearing of pneumonia by the end of the 4th week of therapy. Therefore, repeating the chest X-ray in this patient after 4 *days* is not helpful. In general, chest radiographic clearing of pneumonia is much slower in elderly patients, those who have multi-lobar involvement, those with underlying lung disease, and in alcoholics. ***(Correct Answer: e).***

13) A 33-year-old female is due to undergo abdominal hysterectomy under general anesthesia. She uses daily regular Budesonide/Formoterol inhaler for mild intermittent asthma. The surgeon requests a pre-operative evaluation. She denies nocturnal symptoms, cough, or poor exercise tolerance. She has never been into the Emergency Room before. You don't hear wheezes on chest auscultation. Chest X-ray and spirometry are unremarkable.

What would you tell the surgeon?

a) A 1-week course of prednisolone before the operation

b) Surgery is contraindicated

c) Vaginal hysterectomy using local anesthesia is preferred

d) Anticipate respiratory failure postoperatively and arrange for mechanical ventilation thereafter

e) Proceed with surgery as you have planned

Objective: Review preoperative evaluation of asthma patients.

This patient has well-controlled asthma. Many recent studies have demonstrated that well-controlled asthmatic patients have no significantly increased risk of post-operative pulmonary complications. Therefore, such patients can proceed to surgery at an *average* risk. ***(Correct Answer: e).***

14) A GP is in a dilemma over a case he has referred to you. The patient is a 22-year-old college student who has been receiving treatment for asthma since 2 years in the form of inhaled Fluticasone, Salmeterol, and Albuterol. Her past notes reveal several visits to the Emergency Department for severe symptom worsening. She says she has mild degree of shortness of breath at rest, nocturnal wheeze, and a sensation of neck tightness. You hear a wheeze at the trachea that lessens as you go downward. There are few scattered rhonchi in both sides of the chest. The rest of her examination is, however, unremarkable. You do pulse oximetry, which turns out to be 96%.

What would you do next?

a) Rescue course of oral prednisolone for 2 weeks with gradual tapering

b) Add Montelukast

c) Start Omalizumab

d) Consult the ENT surgeon for laryngoscopy

e) Screen for allergic bronchopulmonary aspergillosis

Objective: Review the differential diagnoses of wheezes.

Superficially, this woman seems to have asthma with frequent exacerbations on the background of prominent daily symptoms. However, her clinical examination suggests vocal cord dysfunction (or paradoxical vocal cord motion) syndrome; there is involuntary closure of the vocal cords, most commonly during inspiration. Most patients complain of episodic and severe shortness of breath with wheezing and intractable cough, often in response to irritant exposures or exercise. This syndrome is more common in women between the ages of 20 to 40 years, and is associated with psychiatric diseases, history of sexual abuse, and gastro-esophageal reflux disease. However, vocal cord dysfunction may co-exist with asthma in 12-56% of cases.

Flexible fiber-optic laryngoscopic examination confirms the diagnosis by showing abnormal adduction of the true vocal cords; this is especially seen in acute episodes. The best step for the time being for this patient is to send her to the ENT surgeon for doing flexible laryngoscope. On doing flow-volume loops, flattening of the inspiratory limb is often detected.

Treatment of acute episodes involves reassurance, panting maneuvers, CPAP, and inhalation of helium-oxygen mixture. Chronic management combines speech therapy, psychological counseling, and avoidance of triggers. ***(Correct Answer: d)***.

15) A 64-year-old man visits the pulmonary outpatients' clinic. He says he has been experiencing progressive shortness of breath with wheezes over the past 5 months. He has mild to moderate COPD for which he receives regular daily inhalers. Eleven months ago, he underwent aortic valve replacement and was admitted to the intensive care unit for prolonged intubation. You examine the patient and do flow-volume loop, which is shown below.

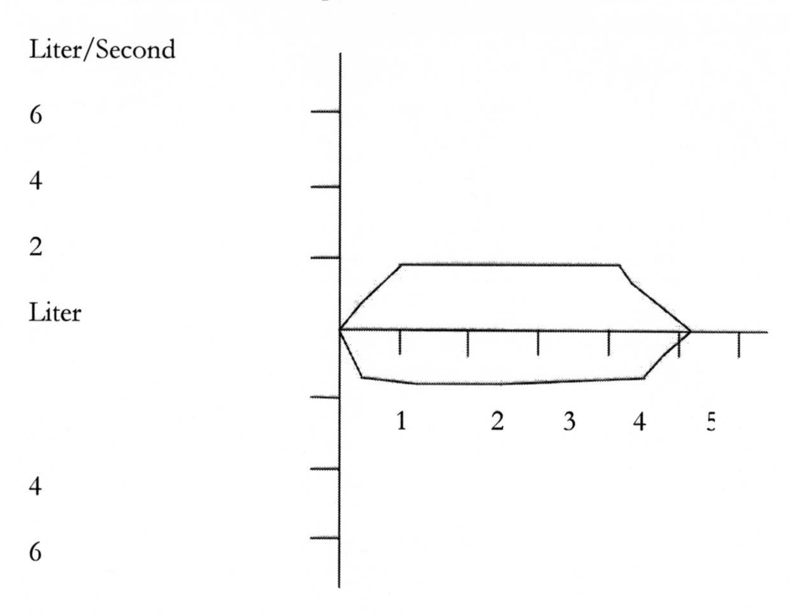

In addition, you do pulmonary function tests, and you get these parameters:

FEV1	2.3 L (65% of predicted
FVC	4.88 L (105% of predicted)
FEV1/FVC	47%
Forced expiratory flow (FEF 25-75%)	2.2 L/sec (60% of predicted)

What is the cause of this patient's *recent* presentation?

a) Left ventricular dysfunction

b) Aortic valve dehiscence

c) Progression of his COPD

d) Co-existent idiopathic pulmonary fibrosis

e) Tracheal stenosis

Objective: Review flow-volume loop as a diagnostic tool in pulmonary diseases.

The patient's pulmonary function tests reflect a moderate degree of COPD. However, the recent, somewhat rapid, progression of symptoms, *flattening* of the inspiratory and expiratory limbs of flow-volume loop (suggesting a *fixed* upper airway obstruction), and the history of prolonged "intubation" all are consistent with tracheal stenosis (which is a well-recognized complication of prolonged intubation). Firm tracheal lesions may limit the modulating effect of transmural pressures on airway luminal diameter. In this pattern, flow is limited during both inspiration and expiration, with flattening of both limbs of the flow-volume loop. FEF(50%)/FIF(50%) ratio close to 1.0 (average 0.9) has been observed as a characteristic of fixed upper airway obstruction. ***(Correct Answer: e).***

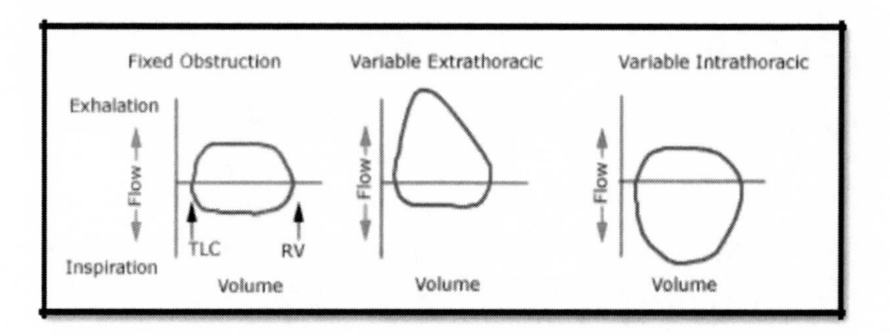

Figure 1. Flow-volume loops in upper airway obstruction. Left panel shows fixed upper airway obstruction with flow limitation and flattening of both the inspiratory and expiratory limbs of the flow-volume loop. Middle panel shows dynamic (or variable, non-fixed) extra-thoracic obstruction with flow limitation and flattening of the inspiratory limb of the loop. Right panel shows dynamic (or variable, non-fixed) intra-thoracic obstruction with flow limitation and flattening of the expiratory limb of the loop. TLC= total lung capacity; RV= residual volume. *From Stoller, JK, Cleve Clin J Med 1992; 59:75.*

Chapter 3

Neurology, Ophthalmology, and Psychiatry
15 Questions

This page was left intentionally blank

1) A 32-year-old high school teacher presents with subacute loss of right eye vision over the past 7 days. She says that her right eye got blurred and the globe movements were painful. She also said that the blurring has gradually increased to a degree that she cannot see anything. She denies headache or orbital trauma. She has been taking oral contraceptive pills over the past 2 years and ibuprofen for occasional back pain. Review of systems is unremarkable, as are her past medical and surgical histories; however, she admits to trying amphetamine 10 years ago, once only. Her older sister has diabetes and her mother has hypertension and chronic glaucoma. Examination of the right eye reveals diminished visual acuity to 6/60, Marcus-Gunn pupil, and normal-looking fundus. The rest of her neurological examination is normal. Brain CT scan is normal but her brain MRI shows many periventricular oval-shaped plaques. CSF examination is unremarkable and her blood tests are within their normal reference range. Urinary and blood toxicology screen is negative. Visual evoked response of the left eye is normal. You start methyl prednisolone infusions for 5 days. She demonstrates remarkable improvement in her right vision within 2 weeks.

How would you manage her in the *long-term*?

a) Daily oral prednisolone

b) Daily azathioprine

c) Monthly methylprednisolone pulses

d) Wait and see

e) Beta interferon therapy

Objective: Review the management of clinically isolated syndromes of demyelination.

This woman presents, for the very first time, with an attack of retrobulbar optic neuritis. The history did not uncover any other neurological complaints (paresthesia, weakness, sphincters, unstable stance…etc.). The investigations are normal apart from the brain MRI which has revealed periventricular plaques. We can diagnose clinically isolated syndrome (CIS) suggestive of multiple sclerosis. About 80% of CIS patients with "abnormal" brain MRI develop the fully-fledged multiple sclerosis after some time.

Approximately, 15-75% of patients with isolated optic neuritis will be diagnosed eventually with multiple sclerosis. There is an accumulating evidence that starting beta interferon (disease-modifying) therapy early in the course of these CISs can improve multiple sclerosis outcome in the long-term (which is definitely better that the wait and see policy). ***(Correct Answer: e).***

2) A 61-year-old retired typewriter presents with an 8-month history of clumsy hands and difficulty walking. He says that he has lost his manual dexterity and has some sort of difficulty using the cutlery and working on the computer. He also mentions that his legs are becoming unable to hold his body when standing or walking. The hands and feet numb all the time and sometimes he feels as if there are worms marching on his feet. He denies any problem with swallowing or control of his urine or bowel. He underwent 2 cataract surgeries last year and he has no double vision. He receives treatment for NYHA functional class II congestive heart failure. His diabetes is unstable for which he has declined the idea of receiving insulin injections. He is compliant with Alendronate tablets, which were prescribed after he had sustained an osteoporotic hip fracture 3 years ago. He neither smokes nor drinks alcohol. His older brother had ischemic stroke. Examination reveals wasting of hands and feet and with diminished pinprick, touch, and joint position sensations. There are proximal and distal weakness, in both upper and lower limbs, with absent deep tendon reflexes. The cranial nerves are intact and there is no sensory level on the trunk. The CSF protein is 200 mg/dl. There is prolonged F-wave latency in the upper and lower limbs in addition to prolonged distal latencies.

What is the best treatment?

a) Prednisolone

b) Plasma exchange

c) Intravenous immunoglobulin

d) Riluzole

e) Interferon beta 1a

Objective: Review treatment options in chronic inflammatory demyelinating polyradiculopathy (CIDP).

Oral prednisolone, plasma exchange, and intravenous immunoglobulin (IVIG) are equally effective in treating CIDP. However, each mode of treatment has pros and cons. His diabetes is unstable and he does not want to take insulin for optimal glycemic control. He has osteoporosis. These would make prednisolone therapy out of question.

Patients with cardiovascular instability are unable to tolerate the fluid shifts incurred by plasma exchange; therefore, this modality can be crossed out. He does not have motor neuron disease (there are prominent sensory signs and symptoms); hence, Riluzole has no place. The overall clinical picture is far away from multiple sclerosis and stem "e" can be cancelled. IVIG is the best treatment option in this man who has co-morbidities. Patients usually start to show improvement after 5 days and this improvement is usually maintained for 2-4 weeks; therefore, repeated administration at 4-week intervals sounds reasonable. IVIG uncommonly can result in headache, renal shutdown, aseptic meningitis, and Coombs positive hemolytic anemia. This form of therapy may affect some lab tests in some way or another transiently:

1. May raise plasma viscosity.
2. May reduce the ESR.
3. May decrease the anion gap (because it has many positively charged components).
4. Antibody titers (ANA, rheumatoid factor, viral antibodies…etc.) cannot be relied upon shortly after the infusion (a positive test may simply be due to the donor's antibodies).
5. Dilutional hyponatremia and pseudo-hyponatremia may develop.
6. Neutropenia can ensue.

(Correct Answer: c).

3) A 42-year-old singer presents with bilateral hand tremor. The tremor has been present since the age of 20 years and is somewhat stable from that time. He denies sustained contractions of the limbs, swallowing difficulty, frequent falls, or jaundice. His father has the same problem. He does not smoke but he does occasionally drink alcohol. Examination reveals postural tremor in both hands. There are normal power, intact deep tendon reflexes, and flexor planters. Sensation is intact as is the sensorium. He takes regular beclomethasone and salmeterol inhalers for asthma. He asks for your help and he is concerned about his embarrassing appearance.

What is the best way to treat this man?

a) Botulinum toxin

b) Reassurance.

c) Propranolol

d) Primidone

e) Alprazolam

Objective: Review treatment options in essential (familial) tremor.

The overall clinical scenario points out towards essential tremor. Either propranolol or primidone can be used as a first-line treatment. This patient's asthma would cancel out propranolol (however, low doses of selective beta blockers, such as atenolol, have been advised in such cases).

Botulinum toxin injection is an excellent option in medically refractory limb, head, and voice tremors. Alprazolam (or clonazepam) are second-line agents, given their physical dependence potential as well as problems with drug withdrawal; these agents are actually widely used because of the wrong belief that hand shaking is due to anxiety. Alcohol has long been shown that it relieves essential tremor, especially essential tremor-related gait ataxia, and drinking small amounts before meals or at social events is actually practiced by many patients who seek temporary relief; however, because of the possibility of misuse and abuse, and the problem of withdrawal, this form of treatment cannot be advised to all patients.

Other agents that can be used are gabapentin and topiramate. That the combination of propranolol and primidone may be better than either alone. ***(Correct Answer: d)***.

4) A 16-year-old high-school boy presents with recurrent syncope. You find bilateral ptosis, impaired eye movements, and pigmentary retinopathy. There is no double vision. His investigations reveal random blood glucose 270 mg/dl; CSF protein 123 mg/dl with 3 mononuclear cells; blood urea 32 mg/dl; Hb 14 g/dl; unremarkable chest X-ray; and normal brain CT scan.

What is the likely cause of this boy's recurrent lapses in consciousness?

a) Generalized seizures

b) Pseudo-seizures

c) Recurrent transient ischemic attacks

d) Uremic encephalopathy

e) Complete heart block

Objective: Review the core features of Kearns-Sayre syndrome.

The presence of bilateral ptosis and ophthalmoplegia but no diplopia in a patient younger than 20 years should always prompt you think of progressive external ophthalmoplegia. This constellation, in addition to diabetes, raised CSF protein, and retinitis pigmentosa, makes Kearns-Sayre syndrome the likely diagnosis. Complete heart block (CHB) is common and may result in recurrent syncope and Stokes-Adams attacks; cardiac involvement is responsible for 20% of causes of death in Kearns-Sayre. All patients should have 12-lead ECG and receive permanent pacemaker if CHB is present. Kearns-Sayre is a mitochondrial cytopathic disease that results from mutations in the mtDNA. Serum lactic acid should be measured, which is elevated, and muscle biopsy examination using modified trichrome staining will reveal ragged red fibers. ***(Correct Answer: e).***

5) A 32-year-old unemployed woman is brought to the Emergency Room by her apartment mate. She is heavily bleeding from a self-inflicted cut in the right forearm. Her apartment mate found her in the bathtub unconscious swamped in her blood. She drinks alcohol and does drugs occasionally and is not taking her Amitriptyline, which was prescribed by her GP. You resuscitate her and she is now stable hemodynamically. She is tearful, has a downcast face, and declines being examined. Her mother has bipolar disorder and her father died of heroin overdose 10 years ago.

What is the best action for the time being?

a) Do toxicology screen.

b) Discharge and give fluoxetine.

c) Cognitive behavioral therapy.

d) Flood therapy.

e) Arrange for electroconvulsive therapy.

Objective: Review the management of failed suicidal attempts.

This young woman has a failed suicidal attempt; there is a very high risk of another attempt if she is allowed to go home. Simply discharging the patient and asking her to take her antidepressant means leaving the patient to her gloomy destiny alone; rapid improvement in the affective symptoms is needed. This can be achieved by using repeated sessions of electroconvulsive therapy. Given the history of drug abuse, screening for HIV, hepatitis C, and hepatitis B viral markers as well as doing toxicology screen would seem reasonable, but these are not urgent steps when a prompt life-protective intervention is required. Flood therapy is part of phobias treatment. Cognitive behavioral therapy has no place in this woman at this time. ***(Correct Answer: e).***

6) A 33-year-old woman visits the multiple sclerosis clinic. She has relapsing-remitting disease. This was diagnosed 2 years when she developed optic neuritis, transverse myelitis, and right-sided trigeminal neuralgia. Over the last year, no relapse was documented. She has been taking beta interferon injections over the past 12 months. She says that she is planning to get pregnant and asks for your help. You find spastic paraparesis, bilateral extensor planters, right-sided primary optic atrophy, right facial numbness, and bi-directional gaze-evoked nystagmus. Brain MRI reveals 9 periventricular oval plaques; one of them took gadolinium avidly.

What would you tell her?

a) Stop interferon beta injections before getting pregnant

b) Pregnancy is contraindicated

c) Continue the current treatment and proceed to pregnancy

d) Take vitamin B6 with interferon injections before getting pregnant

e) You should also take glatiramer acetate with interferon injections during pregnancy

Objective: Review the management of multiple sclerosis during pregnancy.

Multiple sclerosis (MS) patients can get pregnant; the risk of MS relapse is negligible during pregnancy but may increase slightly during early post-partum periods. However, many pregnant patients have difficulty during vaginal labor because of thighs' adductor spasm. Interferons are contraindicated during pregnancy. Although she is maintained in clinical remission, the disease is still active (evidenced by the contrast MRI study; contrast intake indicates active plaques). She has many residual signs. There is no place for combining beta interferons with glatiramer acetate. Neither multiple sclerosis nor its medications affect vitamin B6; replacing this vitamin is not required.

In short, she intends to get pregnant. Although her disease is radiologically active, but it is well-controlled. Simply stop her beta interferon and closely monitor for relapse. If function or life-threatening relapses occur during pregnancy, they can be managed with methyl prednisolone infusions. After pregnancy, resume her beta interferon. ***(Correct Answer: a).***

7) A 62-year-old man is brought to the Emergency Room by his wife. She says that her husband woke up in the morning unable to move his right leg and limb. She also added that she could not understand what he was saying. He has diabetes, hypertension, and chronic tophaceous gout. Examination reveals right-sided hemiplegia, hemi-sensory deficit, and global aphasia. Brain CT scan shows large left hemispheric infarction with mass effect. 12-lead ECG reveals 74 beats/minute sinus rhythm, 30° mean QRS axis, and non-specific ST-T changes.

Which one of the following is *correct*?

a) A search for embolic source should done

b) There is low risk of hemorrhagic transformation when using anticoagulation

c) The left anterior cerebral artery is occluded

d) Hyperglycemia protects the infarcted area

e) Thrombolytic therapy is indicated

Objective: Review the management of acute ischemic stroke.

The main stem of the left middle cerebral artery was blocked occluded. It is a medium-sized artery that is unlikely to be occluded by an *in situ* thrombus formation; a large embolus is almost always the culprit. Echocardiography and carotid artery Doppler studies should be part of the management plan, looking for an embolic source. The infarcted area is large (with mass effect); this argues against the use of thrombolytic therapy (as there is a very high risk of intracranial hemorrhage) while the use of anticoagulation increases risk of hemorrhagic transformation. Hyperglycemia increases neuronal cell death and has an adverse impact in ischemic stroke; near-euglycemia should be achieved by the use of insulin. ***(Correct Answer: a).***

8) A 63-year-old man is brought to Acute and Emergency Department confused. After doing the proper ABC (airway, breathing, and circulation resuscitation), you order some tests. This is his non-contrast brain CT scan.

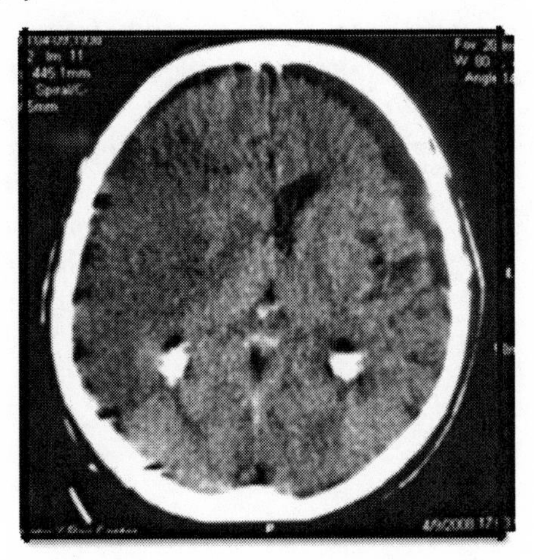

What does the brain CT scan have found? Choose 2:

a) Acute subarachnoid hemorrhage

b) Acute left epidural hematoma

c) Bilateral chronic subdural hematomas

d) Left anterior cerebral artery thrombosis

e) Bilateral basal ganglia calcification

f) Right glioblastoma multiforme

g) Bilateral lobar hemorrhages

h) Right middle cerebral artery occlusion

i) Bilateral frontal lobe atrophy of Picks dementia

j) Giant basilar artery aneurysm

Objective: Review brain CT scan findings of common neurological diseases.

This non-contrast brain CT scan shows a large hypo-dense lesion with mass effect involving the right cerebral hemisphere; in addition, there are bilateral "chronic" subdural hematomas. The former finding is the presenting feature; this is embolic stroke involving the main stem of the right middle cerebral artery. Subdural hematomas are crescent-shaped and compress the underlying cerebral hemispheres

Acute (day 0-3) hematomas appear hyper-dense (fresh blood), while the subacute ones (day 4-21) impart an iso-dense signal with the adjacent brain (because of gradual resolution); chronic hematomas (>21 days) appear hypodense (as in our patient). The bilateral bright signals within the posterior part of the brain represent calcification of the choroid plexuses of the occipital horns of the lateral ventricles; this is neither basal ganglia calcification nor lobar hemorrhage.

Superficially, the frontal lobe appears atrophic (compressed by the subdural hematomas); this may raise a differential diagnosis of frontal lobe atrophy (general paresis of insane, Picks dementia, Huntington's disease…etc.). Acute epidural hematoma appears lens-shaped, extra-axial, hyper-dense mass lesion. The presence of blood at the basal cisterns gives imparts the spider-leg sign of acute subarachnoid hemorrhage. Occlusion of the anterior cerebral artery usually results from *in situ* thrombus formation and produces infarction in the medial frontal lobe. Giant aneurysm of the basilar artery may appear hyper-dense, somewhat rounded, lesion in the posterior fossa (in front of the pons). Glioblastoma multiforme is an aggressive high-grade glioma that typically involves 2 lobes and may extend to the other side via corpus callosum; edema, necrosis, and hemorrhage might be seen. ***(Correct Answers: c, h).***

9) A 32-year-old man presents with rapidly progressive flaccid areflexic paraparesis over the past 8 days. His planter reflexes are flexors. Sensory system examination is unremarkable. You admit him to the neurology ward, and you run a battery of investigations according to your preliminary diagnosis. You get these results:

> CSF: opening pressure 15 cm H_2O; cells 4 (all are mononuclear); protein 200 mg/dl; sugar 78 mg/dl; Gram stain negative; Ziehl-Neelsen stain negative.
>
> Blood sugar 100 mg/d; WBCs 5,300/mm^3; hemoglobin 13.5 g/dl; ESR 12 mm/hour; stool examination no blood, pus, or parasites.

All of the following with regard to this man's illness are incorrect, *except*:

a) Shortness of breath may be due to interstitial lung fibrosis

b) Tachyarrhythmia mainly result from myocardial involvement

c) Preceding *Campylobacter jejuni* infection portends a poor prognosis

d) Extra-ocular muscle weakness is common

e) Most signs are asymmetrical at the start of the illness

Objective: Review acute demyelinating radiculopathies.

Summarize what you have concluded from this scenario. This man has developed rapidly progressive flaccid paraparesis normal sensation, and no pyramidal signs. His bloods are unremarkable but his CSF shows albumin-cytological dissociation. This would fit Guillain-Barre syndrome. Those patients may develop shortness of breath because of respiratory muscle weakness, pulmonary thromboembolism, or aspiration pneumonia (because of bulbar weakness). The disease does not incite an interstitial lung fibrotic reaction (neither in the short-term nor in long-term course). Dysautonomia is common, and fluctuation in blood pressure and heart rate occurs; cardiac failure is not part of this syndrome. Facial weakness is noticed in 50% of patients, while extra-ocular muscle paresis is very unusual.

Some degree of asymmetry in the lower limbs' weakness may be found in 9% of patients at the presentation; however, striking asymmetry or persistent asymmetry in the signs should cast a strong doubt on the diagnosis of Guillain-Barre syndrome. A preceding *Campylobacter jejuni* infection portends a poor prognosis; axonopathy, need for assistant ventilation, and rapid downhill course are the other poor prognostic indicators. ***(Correct Answer: c).***

10) A 27-year-old pregnant woman presents with a 2-day history of confusion and double vision. It is the 3rd trimester. You find fever, neck stiffness, and absent right corneal reflex. Brain CT scan is unremarkable. You proceed to lumbar puncture. The CSF profile is as follows:

Opening pressure: 24 cm H_2O

Cells: 132/mm^3, 72% neutrophils and 28% lymphocytes

Protein: 110 mg/dl

Sugar: 32 mg/dl (blood sugar 100 mg/dl)

What does the woman have?

a) Pneumococcal meningitis

b) *Herpes simplex* encephalitis

c) Burst brain abscess

d) *Listeria monocytogenes* meningoencephalitis

e) Mollaret meningitis

Objective: Review characteristic features of various CNS infections.

It is obvious that this woman suffers an acute meningitic illness. The prominent brainstem signs in this pregnant woman would suggest *Listeria* meningoencephalitis (*Listeria* prefers to target the brainstem). Her CSF profile is also compatible with this infection. A burst brain abscess would be seen on the brain CT scan and the resulting meningitis would result in a "very" high CSF neutrophilic pleocytosis. *Herpes simplex* encephalitis attacks the fronto-temporal areas of the brain. Mollaret meningitis is a form of aseptic meningitis; the sugar is not reduced. The prominent brainstem signs would favor stem "d" over stem "a". ***(Correct Answer: d).***

11) A 6-year-old child is brought by his parents to consult you. After you interview and examine the patient, you conduct some tests and you get this:

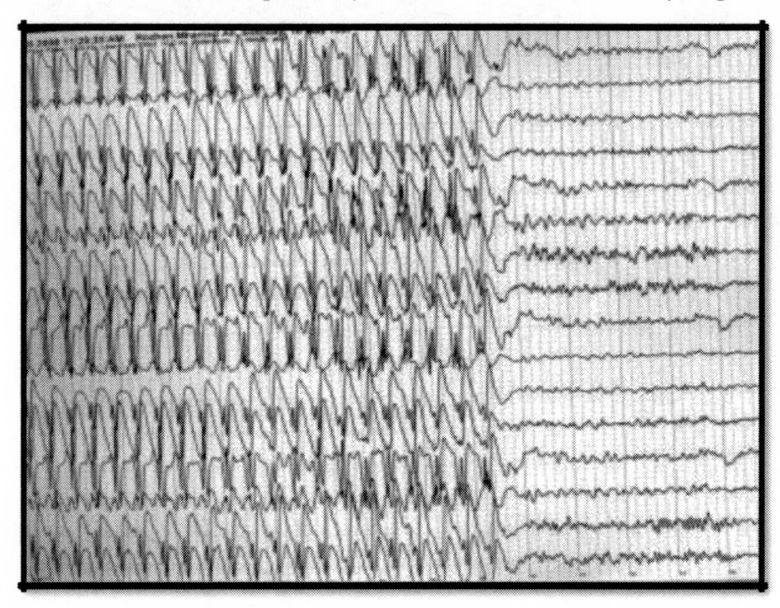

Which one of the following is the *correct* statement with respect to this child's illness?

a) Salaam attacks is the diagnosis

b) Carbamazepine is contraindicated

c) Lamotrigine is useless

d) Hippocampal sclerosis is the underlying etiology

e) Never associated with juvenile myoclonic epilepsy

Objective: Review characteristic EEG findings in various seizure disorders.

This is an on-spot diagnosis; generalized 3 Hz spike-and-wave discharges. This is absence epilepsy. The scenario did not give a single clue. Phenytoin and carbamazepine worsen this type of epilepsy; valproate, lamotrigine, clonazepam, and ethosuximide represent the most effective therapies.

Absence epilepsy may be associated with juvenile myoclonic epilepsy. The prognosis for seizure remission differs for each syndrome.

Childhood absence epilepsy (typical absence seizures, occasional generalized tonic-clonic seizures, EEG pattern of generalized 3-Hz spike-and-wave activity in a neurologically healthy child) has a reasonably good long-term prognosis with disappearance of seizures in the teen years; juvenile myoclonic epilepsy is associated with life-long seizures, however.

Hippocampal sclerosis has been linked to temporal lobe epilepsy syndrome, and salaam attacks (infantile spasms) typically start in the 1st year if life (and their EEGs show hypsarrhythmia).

(Correct Answer: b).

12) A 54-year-old man is brought to the Emergency Room unconscious. He has long-standing hypertension and diabetes, both of which are well-controlled with Enalapril and Metformin, respectively. His Glasgow Coma Scale is 5/15. Urgent non-contrast brain CT scan shows this:

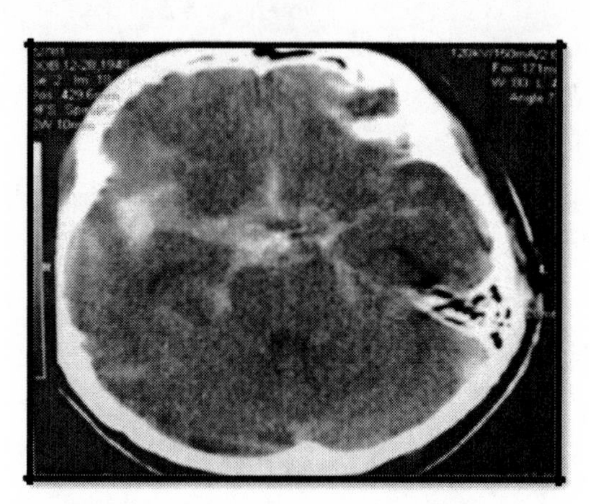

What is the *correct* statement regarding this man's disease?

a) Conventional 4-vessel cerebral angiography should be arranged

b) This presentation is a consequence of his diabetes

c) CSF examination is very helpful to consolidate the diagnosis

d) Gradual clinical improvement is expected within the next 2 weeks

e) Early onset seizures predict a poor outcome

Objective: Review the management of acute subarachnoid hemorrhage.

The brain imagining shows blood filling in the basal cisterns in a spider-leg pattern, that is suggestive of acute subarachnoid hemorrhage (SAH). Most of the blood appears to be within the right Sylvian fissure, locating the ruptured aneurysm to the right middle cerebral artery. The Glasgow Coma Scale (GCS) is not a true SAH grading scale but is rather a standardized scale for evaluating the level of consciousness.

However, it is widely known and has some utility for predicting outcome after SAH. Sedating medications and intubation can confound interpretation of the GCS.

This patient has a GCS of 5/15, implying a profound loss of consciousness and poor outcome; such patients are poor candidates to surgical or endovascular intervention, and therefore, angiographic evaluation is not recommended. Diabetes mellitus is not a risk factor for SAH; smoking, hypertension, alcohol abuse, and positive family history are well-established risk factors for SAH. Seizures at the onset of SAH appear to be an independent risk factor for late seizures and a predictor of poor outcome. The incidence of late epilepsy after surgical management of SAH is unclear. The diagnosis in this man is clear-cut, and there is no justification for doing lumbar puncture. ***(Correct Answer: e).***

13) A 26-year-old woman presents with a 1-week history of pancephalic headache, paraparesis, and focal seizures, few days after giving birth to a full-term healthy-looking male baby. Her initial brain CT scan was unremarkable. You do brain MRI without gadolinium and you notice this:

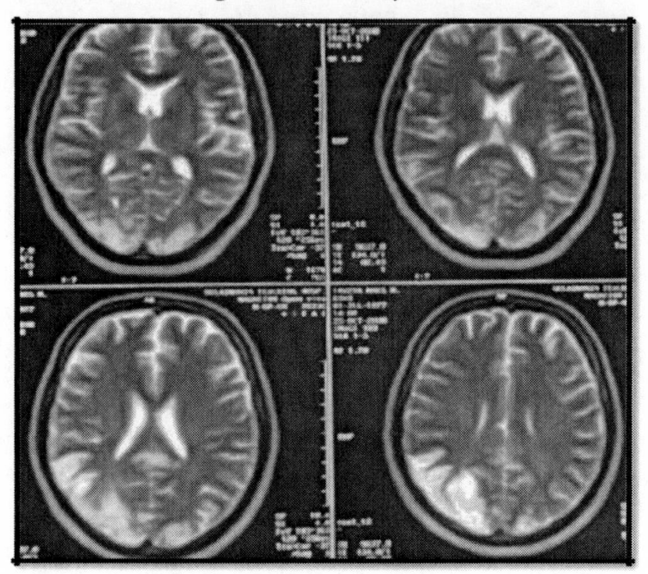

What is the cause of this woman's presentation?

a) Acute hydrocephalus

b) Cerebellar hemorrhage

c) Right posterior cerebral artery occlusion

d) Ruptured basilar aneurysm

e) Superior sagittal sinus thrombosis

Objective: Review the diagnostic approach of intracranial venous sinus thromboses.

The occurrence of headache, paraparesis (due to damage to both leg's cortical areas), and focal seizures (the hemorrhagic venous infarcts are highly epileptogenic) over one week, in addition to normal initial brain CT scan should always prompt the physician search for superior sagittal sinus venous thrombosis. Pregnancy and puerperium are well-known hypercoagulation situations.

Symptoms and signs of cerebral venous sinus thrombosis can be grouped in three major syndromes; isolated intracranial hypertension syndrome (headache with or without vomiting, papilloedema, and visual problems); focal syndrome (focal deficits, seizures, or both); and encephalopathy (multifocal signs, mental status changes, stupor or coma).

With superior sagittal sinus occlusion, bilateral motor deficits and seizures are frequent, while presentation as an isolated intracranial hypertension syndrome is infrequent. Patients with isolated lateral sinus thrombosis present mostly with isolated intracranial hypertension; however, if the left transverse sinus is occluded, aphasia often follows. Head CT scan is normal in up to 30% of cerebral venous sinus thrombosis, and most of the findings are actually nonspecific. However, in about one-third of patients, CT demonstrates direct signs of cerebral venous sinus thrombosis, which are the empty delta sign, the cord sign, and the dense triangle sign. This patient's T_2-weighted MRI film shows bilateral asymmetrical hyper-intense signals at both parieto-occipital areas with sulcal effacement; these are hemorrhagic venous infarcts with surrounding cytotoxic edema. Hemorrhagic venous infarcts appear hyper-intense on both MRI sequences. ***(Correct Answer: e)***.

14) A 61-year-old man presents with pancephalic headache, protracted vomiting, and confusion over 4 weeks. He has left-sided pyramidal weakness and papilloedema. Neither neck stiffness nor fever is detected. These are his brain MRI images:

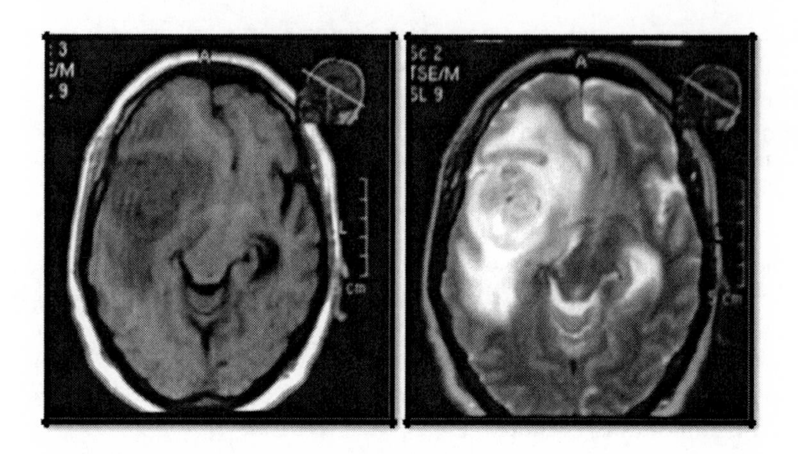

What does the man have?

a) *Herpes simplex* encephalitis

b) Right basal ganglia hemorrhage

c) Right middle cerebral artery occlusion

d) Glioblastoma multiforme

e) Right hemispheric meningioma

Objective: Review the brain MRI findings of various gliomas.

This abbreviated scenario points out towards a rapidly expanding intracranial space-occupying lesion. The brain MRI images (T_1 and T_2-weighted, left and right respectively) show a large mass in the right fronto-temporo-parietal areas with massive surrounding edema and midline shift. The multi-lobar involvement is highly characteristic of high-grade gliomas.

These tumors account for about 20% of all intracranial tumors, or about 55% of all tumors of the glioma group, and for more than 90% of gliomas of the cerebral hemispheres in adults. The tumor has a variegated appearance, being a mottled gray, red, orange, or brown, depending on the degree of necrosis and presence of hemorrhage, recent or old.

The imaging appearance is usually that of a non-homogeneous mass, often with a center that is hypo-intense in comparison to adjacent brain and demonstrating an irregular thick or thin ring of enhancement, and surrounded by edema. Part of one lateral ventricle is often distorted, and both lateral and third ventricles may be displaced contralaterally. ***(Correct Answer: d).***

15) A 71-year-old man is brought by his family to consult you. The man has been experiencing easy forgetfulness over the past few months. You interview the patient and examine him. He scores 22 on the mini-mental status examination. Brain CT scan reveals diffuse cortical atrophy and hydrocephalus. You diagnose Alzheimer's disease.

Which one of the following would you use?

a) Rivastigmine

b) Memantine

c) Vitamin E

d) Stem cell transplantation

e) Ventriculo-peritoneal (VP) shunt

Objective: Review management of Alzheimer's disease.

The patient's mini-mental status examination score is 22. This is mild Alzheimer's disease; Memantine is used in the treatment of moderate-severe disease. The hydrocephalus compensatory and is secondary to global brain atrophy; no need for VP shunting. Rivastigime, Donepezil, and Galantamine are central acetylcholinesterase inhibitors that are used in the treatment of mild cases. ***(Correct Answer: a).***

Chapter 4

Pharmacology, Therapeutics, and Toxicology

10 Questions

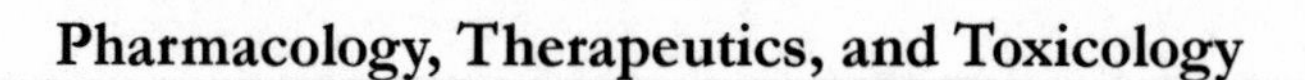

This page was left intentionally blank

1) A 43-year-old man underwent allogenic renal transplantation 4 weeks ago, because of end-stage renal disease of unknown cause. He takes daily cyclosporine, azathioprine, and prednisolone since then. He was doing well until this morning. Today, he developed 3 generalized tonic-clonic seizures. His GP has been treating him for flu using clarithromycin and paracetamol. His flu seems to be improved. The patient's wife denied head trauma and she says that her husband is fully compliant with his daily medications. His father died at the age of 63 years because of diabetic nephropathy and his older brother has idiopathic generalized epilepsy. The patient's past records uncover pyogenic meningitis at the age of 6 years and tonsillectomy at the age of 8 years. Examination reveals no fever or neck stiffness. The patient is fully conscious. There are bilateral flexor planter reflexes. Blood urea and serum creatinine have been stable over the past 2 weeks.

What is the likely cause of these seizures?

a) Cerebral toxoplasmosis

b) *Streptococcal* meningitis

c) Idiopathic grand mal epilepsy

d) Cyclosporine toxicity

e) Uremic encephalopathy

Objective: Review cyclosporine interactions.

This man has been doing well since the time of transplantation. A new medication has been used recently (to treat flu). The family history is irrelevant to the new presentation; a distraction! Past medical and surgical histories have nothing to do with this new presentation, as well. Cyclosporine is metabolized by the hepatic cytochrome P450A3 system, which is affected by many medications and drugs.

The following *increase* cyclosporine blood concentrations: diltiazem, nicardipine, verapamil, fluconazole, itraconazole, ketoconazole, clarithromycin, erythromycin, lansoprazole, rabeprazole, cimetidine, methylprednisolone, allopurinol, bromocriptine, metoclopramide, colchicine, amiodarone, danazole, and grapefruit juice.

The following *lower* the cyclosporine blood concentrations: nafcillin, rifabutin, rifampicin, carbamazepine, phenobarbital, phenytoin, octreotide, ticlopidine, orlistat, and St. John's Wort.

Neurological complications are usually reversible upon lowering the daily drug dosage (or stopping it briefly if the condition permits) or changing the intravenous form to oral formulations. The neurotoxicity of cyclosporine may manifest as severe headache, visual disturbances, and seizures; this syndrome is associated with acute hypertension, and resembles hypertensive encephalopathy (posterior leukoencephalopathy is usually seen on brain imaging). Mild tremor is common with therapeutic doses of cyclosporine, occurring in 35 to 55% of patients (it may improve despite continued therapy). ***(Correct Answer: d).***

2) A 41-year-old man has been referred to your office from the psychiatry department. The referral letter states that the patient receives medical therapy for chronic schizophrenia but he has developed jaundice 2 weeks ago. The patient says that he is compliant with his daily chlorpromazine tablets. He also reports itching. He is unsure if there are any changes in the color of urine or stool. His schizophrenia has been fluctuating over the last 10 years. He denies nausea, vomiting, abdominal pain, or any change in his bowel habit. He drinks 2 cans of beer every other day and smokes 1 cigar daily. He displays no risk factors for HIV infection. His mother has affective psychosis. Examination reveals yellowish discoloration of the sclera, generalized hyperpigmentation, and excoriation marks. Hepatitides screen reveals positive serum IgG anti-HBs. Total serum bilirubin is 4 mg/dl, serum indirect bilirubin is 1 mg/dl, serum alkaline phosphatase is 200 iu/L, and serum AST and ALT are 35 iu/L and 34 iu/L, respectively.

What is the reason behind his current complaint?

a) Alcoholic hepatitis

b) Non-alcoholic hepatic macrosteatosis

c) Chlorpromazine-induced cholestatic jaundice

d) Chronic hepatitis B infection

e) Primary biliary cirrhosis

Objective: Review the complications of chlorpromazine therapy.

This man has developed jaundice, itching, and lab evidence of cholestatic hepatitis; stem "c" would fit this constellation. Generalized hyperpigmentation is due to chlorpromazine therapy and is not due to primary biliary cirrhosis (this patient has no clubbing; the history is short; the question did not address positive anti-mitochondrial antibodies).

Non-alcoholic hepatic macrosteatosis typically displays normal or slightly raised serum transaminases (usually ALT); jaundice and cholestasis are not features of this disease. The drinking history does not refer to a possible development of alcoholic hepatitis; the lab findings are inconsistent.

The presence of positive IgG anti-HBs antibodies (only; no other viral markers were found) indicates past vaccination (although the patient did not tell us about this); positive serum testing for HBs antigen with IgG anti-HBc antibodies are seen in chronic hepatitis B infection. ***(Correct Answer: c).***

3) A 56-year-old woman is brought to the Emergency Room by her brother. The patient is unconscious, not responding to verbal or painful stimuli. She has rapid deep breathing. He found her on her bed with 3 empty containers of adult aspirin. The husband says that she been extremely anxious recently because of their divorce plan. You do the ABC (airway, breathing, and circulation) and arrange for blood tests. Chest auscultation is normal.

What else you will do?

a) Brain CT scan

b) Chest X-ray

c) Hemodialysis

d) Blood methanol and ethylene glycol

e) Blood urea and electrolytes

Objective: Review the management of aspirin poisoning.

Despite extensive clinical experience with aspirin and other salicylates, and the evolution of therapeutic alternatives, salicylate intoxication remains a significant clinical problem. Early recognition is the key to successful management, and this diagnosis must be considered in any patient following therapeutic drug overdose and in those with an unexplained increase in the anion gap. The presence of any one of the following should call for hemodialysis in acute aspirin poisoning: coma, seizures, pulmonary edema, renal impairment, and aspirin blood level >400 mg/dl in young patients (or >300 mg/dl in elderly patients). Aspirin poisoning results in combined metabolic acidosis and respiratory alkalosis. In mild poisoning, ABC resuscitation, forced alkaline diuresis, and close monitoring would suffice.

The causes of suicide are depression, schizophrenia, drug and alcohol abuse, and acute stressful life events. There are no chest signs; chest X-ray is not needed. Brain CT scan will reveal no changes in aspirin poisoning. ***(Correct Answer: c)***.

4) A 44-year-old alcoholic man is brought to the Emergency Room. He is confused and has developed generalized tonic-clonic seizures over the past few hours. The medical lab has performed some blood and urine tests. These are:

Serum sodium	137 mEq/L
Serum potassium	3.7 mEq/L
Serum chloride	102 mEq/L
Serum bicarbonate	10 mEq/L
Blood urea	56 mg/dl
Serum creatinine	1.6 mg/dl
Serum glucose	90 mg/dl
Urine examination	RBCs 10/HPF, full of oxalate crystals
Plasma osmolality	325 mosm/Kg

What does the man have developed?

a) Alcoholic blackout

b) Alcohol withdrawal rum fits

c) Alcohol intoxication

d) Ethylene glycol poisoning

e) Alcoholic ketoacidosis

Objective: Review ethylene glycol poisoning.

Interpret the metabolic panel this man has. There are high anion gap metabolic acidosis, mild renal impairment, increased plasma osmolality (and osmolal gap), and bloody urine that is full in oxalate crystals.
Some alcoholic individuals, when out of alcohol, crave for alternatives; methanol or ethylene glycol types of alcohol are the usual substitutes. The parent alcohol, ethylene glycol, is relatively nontoxic, and mainly results in central nervous system sedation if ingested; its metabolites (glycolate, glyoxalate, and oxalate) are highly toxic to many organ systems. Ethanol or methanol may also be co-ingested.

The diagnosis needs a high index of suspicion and vigilance; serum methanol and ethylene glycol concentrations are usually performed by gas chromatography, but they are not widely available and frequently must be sent to a reference laboratory. Such send-out laboratory testing rarely, if ever, returns in time to assist clinical decision-making. The occurrence of hematuria and oxalate-full urine in the appropriate clinical setting suggest ethylene glycol poisoning, while the development of visual impairment, scotomas, and even total blindness should prompt you think of methanol intoxication. This man's mild azotemia, hematuria, and oxalate crystal-full urine are indicative of ethylene glycol poisoning. ***(Correct Answer: d)***.

5) 25-year-old man presented to his GP with sore throat and enlarged tender cervical lymph nodes, few days ago. Oral ampicillin was given and the patient's pharyngitis started to resolve. Today, he is visiting the physician's office because of fever, skin rash, and joint pains. His past medical history is notable for gonorrhea, 2 years ago, and typhoid fever, 9 months ago. There is no family history of note. He denies chest pain, bowel symptoms, or dysuria, and he says that his urine output is maintained. Blood tests reveal leukocytosis and eosinophil count of 10%, mildly raised blood urea and serum creatinine. Urine is positive for protein and blood, and contains many eosinophils.

What is the cause of the new presentation?

a) Leptospirosis.

b) Disseminated gonococcal disease.

c) Epstein-Barr viral infection.

d) Hanta viral infection.

e) Drug-induced acute allergic interstitial nephritis.

Objective: Review adverse drug reactions.

This man's illness started with bacterial pharyngitis, which was responsive to oral ampicillin therapy. However, and later on, he started to develop fever, arthralgia, and skin rash in addition to non-oliguric acute renal impairment, eosinophilia and eosinophiluria. Drug-induced acute allergic interstitial nephritis would fit this clinical scenario. The history of gonorrhea is unrelated to his current presentation and stem "b" is a distraction. Stem "a", as in post-streptococcal glomerulonephritis, presents with rapidly progressive renal failure and diminished urine output, edema, hypertension, and active urinary sediment (with casts). Stem "c" is a possibility, because of the appearance of skin rash after starting ampicillin. However, the virus rarely attacks the kidneys in such a degree, and does not produce eosinophilia. Hanta virus can attack the lungs and kidneys but it has a very different clinical picture (which may include hemorrhagic fever presentation). ***(Correct Answer: e).***

6) A 31-year-old widow is brought to Emergency Department by her older sister. The patient has global confused, tachycardia, and flushing. Her sister says that the patient has been receiving a medication for low mood treatment since the death of her husband, 3 months ago. Two hours ago, she found her on her bed irritable; many tablets of a medication were seen on the floor. Examination reveals blood pressure of 80/40 mmHg, flushing, dry skin, dilated pupils, and hyperactive reflexes. Resuscitation has begun with decontamination. ECG shows regular heart rate of 140 beats/minute and QRS complex duration of 160 msec. Normal saline and sodium bicarbonate infusions are being given. Two hours later, the blood pressure becomes 80/55 mmHg and the QRS complex duration becomes 120 msec.

Choose the *correct* action that should be undertaken for the time being?

a) Mechanical ventilation

b) Arrange for hemodialysis

c) Give epicac

d) Infuse magnesium

e) Noradrenalin infusion

Objective: Review the management of tricyclics poisoning.

Although selective serotonin reuptake inhibitors have become the first-line agents for the treatment of depressive disorders, tricyclics are still used for depressed patients as well as in chronic pain syndromes, obsessive-compulsive disorders, and panic disorders; clinical intoxication (whether intentional or accidental) is still seen. Tricyclic antidepressants (TCAs) intoxication usually produces sinus tachycardia and hypotension; ventricular tachycardia and fibrillation are seen in only 4% of cases. Actually, refractory hypotension is responsible for the majority of TCA-related mortality. Bicarbonate infusion usually improves hypotension (and shortens the prolonged QRS duration) and crystalloids have an additive effect. However, when the blood pressure fails to rise, intravenous vasopressors should be administered (such as noradrenalin in this patient). Sodium channel blockers (e.g., phenytoin) are ineffective in the treatment of toxin-medicated seizures (benzodiazepines are the agents of choice to increase the brain's GABAergic transmission).

TCAs are not dialyzable, as they are tightly protein-bound. Although severe and advanced cases can have apnea, the question did not address any respiratory compromise and stem “a” is not applicable.

Her ECG did not show polymorphic ventricular tachycardia and stem “d” cannot be given as a prophylaxis. Epicac is no longer recommended for upper GIT decontamination. ***(Correct Answer: e).***

7) A 19-year-old female has been referred from a rural hospital. The referral letter states that the patient had ingested many tablets of paracetamol but it mentions neither the timing of ingestion nor the likely ingested dose. You check serum paracetamol and it turns out to be 19 μg/ml. You consider role of *N*-acetylcysteine as an antidote in this patient.

How would you reply?

a) *N*-acetylcysteine should be given after 24 hours

b) *N*-acetylcysteine is contraindicated

c) *N*-acetylcysteine should be given after checking liver function tests

d) *N*-acetylcysteine is indicated and should be given now

e) *N*-acetylcysteine should be given after consulting the liver transplant unit

Objective: Recall the treatment of paracetamol poisoning and the indications of N-acetylcysteine (NAC).

NAC is indicated in the following situations:

1. Patients with serum acetaminophen concentration above the "possible hepatic toxicity" line of the Rumack-Matthew nomogram following an acute ingestion.

2. Patients with a single ingestion of >150 mg/kg (or 7.5 g in an adult) by history and for whom results of serum acetaminophen concentration will not be available within 8 hours from the time of ingestion.

3. Patients with an unknown time of ingestion and serum acetaminophen concentration >10 μg/mL (*as in our patient*).

4. Patients with laboratory evidence of hepatotoxicity (from mildly elevated aminotransferases to fulminant hepatic failure) and a history of excessive acetaminophen ingestion.

5. Patients who have ingested repeated excessive acetaminophen doses, have risk factors for acetaminophen-induced hepatotoxicity, and a serum acetaminophen concentration >10 μg/mL.

(Correct Answer: d).

8) A 43-year-old man visits the diabetic outpatient clinic for a scheduled follow-up. He has long-standing type I diabetes and hypertension. He also has prostatic enlargement, chronic plaque psoriasis, diabetic background retinopathy, and diabetic autonomic gastroparesis. The diabetologist considers the use of pramlintide for optimal glycemic control. However, you disagree with the diabetologist and think that pramlintide is contraindicated in this man.

Why?

a) Because of his hypertension.

b) Because of his prostatism.

c) Because he has type I, not type II, diabetes.

d) Because of his gastroparesis.

e) Because of his psoriasis.

Objective: Recognize the role of amylin analogs in the treatment of diabetes.

Pramlintide is an amylin analog and is only approved for use in patients who are taking insulin. However, it requires injections with each meal in the setting of type I diabetes, and at least twice daily injections in type II diabetes. It cannot be mixed in a syringe with insulin. Its contraindications are hypersensitivity to pramlintide or any component of the formulation; confirmed diagnosis of gastroparesis; and hypoglycemia unawareness. ***(Correct Answer: d).***

9) A 23-year-old hemophiliac A man consults about his treatment. He says that his factor VIII concentrate has not been that effective recently in stopping his bleedings. You check his serum factor VIII inhibitors and the result is 18 BU. You explain to him that there are certain antibodies in his blood and these are directed against the infused factor VIII concentrate, rendering the infusions virtually useless. You educated him about the use of activated prothrombin complex concentrates (factor VIII inhibitor bypass activity; FEIBA).

Which one of the following you have mentioned to him about this medication?

a) It is safe in hepatic impairment

b) It can result in thrombotic complications

c) Infusion-related reactions don't occur

d) It can be given as intramuscular injections

e) It is used in hemophilia A, but not hemophilia B, patients

Objective: Recall the management of factor VIII concentrate inhibitors.

Generally, activated prothrombin complex concentrate is expensive, provides unpredictable hemostasis without the ability to monitor clinical efficacy with a laboratory test, and carries the risk of significant complications. Extreme caution is required in the presence of hepatic impairment. Thrombotic complications may occur, including DIC and myocardial infarction. Only the intravenous route is used. Infusion-related reactions with fever and chills may develop. It can be given in hemophilia A and B patients who have significant plasma titer of factor VIII inhibitors. ***(Correct Answer: b).***

10) A 58-year-old man has been receiving Rituximab infusions for rheumatoid arthritis since 9 months. However, he presents today with slowly progressive hemiparesis and visual field defects, which have been progressing over the past 2 months. Non-contrast brain CT scan is unremarkable.

What's your preliminary diagnosis?

a) Brain abscess

b) Malignant hemispheric glioma

c) Progressive multifocal leukoencephalopathy

d) Transverse myelitis

e) Multiple sclerosis

Objective: Recognize that Rituximab is associated with the development of progressive multifocal leukoencephalopathy (PML).

PML has previously been associated with Rituximab use, although previous reports were confined to patients treated for hematologic malignancies and an unapproved use, systemic lupus erythematosus. However, it is now established that patients with rheumatoid arthritis who receive this medication can also develop PML. ***(Correct Answer: c)***.

This page was left intentionally blank

Chapter 5

Hematology, Oncology, and Palliative Medicine

15 Questions

This page was left intentionally blank

1) A 58-year-old carpenter presents with fatigue for 6 months. He says that he is unable to do his daily job, as he has no power to work and this has forced him to supervise the work only. His wife says that he eats and sleeps well but he is obviously pale. He denies any change in his bowel habit. There is no weight loss but he has dyspepsia, which was ascribed to irritable bowel syndrome, as his GP told him. He has stage I hypertension and takes daily hydrochlorothiazide. He takes occasional anti-spasmodic medications for irritable bowel. His father died of colonic cancer at the age of 70 years and his mother died of massive stroke at the age of 75 years. His younger sister has Crohn's disease. He smokes 2 to 4 cigarettes per day and drinks a glass of wine every night. His GP has ordered colonoscopy for him, which turned out to be normal. Hemoglobin is 6.7 g/dl, RBCs are hypochromic microcytic, but the WBCs and platelets are normal. Liver transaminases, blood urea and electrolytes are within their normal reference range. Serum ferritin is 40 μmol/L.

What would you do to discover the cause of his fatigue?

a) Repeat colonoscopy

b) Refer for upper GIT endoscopy

c) Do sigmoidoscopy

d) Order fecal occult blood testing

e) Bone marrow study

Objective: Review the diagnostic approach of iron deficiency anemia.

This man's clinical picture of hypochromic microcytic anemia and dyspeptic symptoms should always call for searching a GIT source of blood loss. His family history demonstrates 2 GIT diseases; one of them is a GIT malignancy. Although he has no bowel habit change, his GP was correct in ordering colonoscopy, the result of which was normal, however. A good step to do now is to proceed with upper GIT endoscopy. Fecal occult blood testing is neither sensitive nor specific for diagnosing GIT malignancies, and the test can be truly negative while the patient has a genuine GIT blood loss. Repeating colonoscopy or doing sigmoidoscopy would be unreasonable. Bone marrow study would confirm iron deficiency anemia but would not uncover its cause. (***Correct Answer: b.***).

2) A 55-year-old office manager comes for his annual check-up visit. He is reasonably well and healthy and has no chronic diseases. He lives alone in an apartment, does the shopping, and jogs every Sunday morning in the local park. He does not drink but does smoke a cigarette or two every day. His past medical and surgical histories are unremarkable, as is his family history. Examination reveals no organomegaly or lymph node enlargement. Abdominal ultrasound did not show liver, spleen, or lymph node enlargement. His complete blood count shows hemoglobin 13 g/dl, white cells 25 x 10^9/L with 95% mature-looking small lymphocytes, and platelets count 200 x 10^9/L.

Which one of the following is true regarding this man's disease?

a) He has stage I disease

b) He should receive monthly intravenous immunoglobulin infusion

c) The best approach is observation

d) There is no increased risk of developing solid malignancy

e) Short doubling time carries good prognosis

Objective: Review the treatment options in chronic lymphocytic leukemia (CLL).

About 25% of CLL cases are diagnosed incidentally by performing blood counts for some reason or another. Lymphocytosis (with small mature-looking lymphocytes and smudge cells) is the first clue in such cases. This man has no enlarged lymph nodes or organomegaly. This would categorize him as stage 0 (according to the Rai's staging system) which confers a median survival of 150 months; therefore, observation is all that is required. Patients with repeated major infections (septicemia, severe pneumonia…etc.) should receive monthly intravenous immunoglobulin infusions. Short doubling time portends a bad prognosis. Around 10% of CLL patients will develop Richter's transformation, and another 10% will proceed to prolymphocytoid transformation. Transformation to acute leukemia is unusual. However, CLL patients have greater than average risk of developing solid malignancy (usually of the lung, GIT, and skin). ***(Correct Answer: c).***

3) A 28-year-old woman is referred to the general medical ward with a 3-day history of progressive pallor and a tinge of jaundice. She denies bowel symptoms, fever, or bleeding. She was admitted to the hospital 12 days ago, and treated in the orthopedic department for fractured left mid-femur. She lives with her husband and has 2 healthy daughters. She takes no medications or illicit drugs but drinks alcohol occasionally. Hb 7 g/dl, reticulocytes 6%, no bilirubin in urine, blood urea 31 mg/dl, and ALT 12 iu/L.

What is the likely cause of this woman's presentation?

a) Acute acalculous cholecystitis

b) Rhabdomyolysis

c) Delayed hemolytic transfusion reaction

d) Viral hepatitis C

e) Transfusion of old stored blood

Objective: Review the diagnostic approach of transfusion-associated jaundice.

She developed a fracture in her right femur. Such fractures are notorious for causing bleeding; she must have been given blood transfusions. Massive old blood transfusion can result in jaundice in the early postoperative period but will not explain the occurrence of this progressive pallor and jaundice after 1 week. The incubation period of hepatitis C is long and is unlikely to be the cause of this pallor. Severe crushing injuries may cause rhabdomyolysis 1-2 days after the trauma but not after 1 week. The absence of abdominal symptoms (pain, nausea) argues against the development of cholecystitis; in addition, it does not explain the low hemoglobin. Delayed hemolytic transfusion reaction is the only explanation. Note the low hemoglobin, gross reticulocytosis, and acholuric jaundice 1 week after the trauma; she has 2 children and might well be sensitized to certain RBC antigens (often of the Kidd or Rh system) during her past pregnancies; with the recent RBCs antigen re-exposure, she has developed this type of reaction. A typical patient would demonstrate a falling hematocrit, low grade fever, mild increase in serum indirect bilirubin, and spherocytosis on the blood smear may be found in association with a new positive direct antiglobulin (Coombs) test and a new positive antibody screen. ***(Correct Answer: c).***

4) A 34-year-old man is referred to you from the gastroenterology clinic. He has chronic diarrhea, which was ascribed to selective IgA deficiency. He has chronic sinusitis for which he takes certain nasal drops and tablets. Examination fails to detect any organomegaly, lymphadenopathy, pallor, or clubbing. One of his family members has common variable immune deficiency syndrome.

Which one of the following is the best life-long treatment of this man's immune deficiency state?

a) Monthly intravenous immunoglobulin infusions

b) Bi-monthly IgA infusions

c) Bi-weekly fresh frozen plasma infusions

d) No specific treatment is available

e) Bone marrow transplantation

Objective: Review the management of selective IgA deficiency.

Selective IgA deficiency has no specific treatment. There is no IgA preparation to be infused. Some of these patients have antibodies against IgA molecules; blood or blood product transfusion may result in fatal anaphylactic reactions. Symptomatic measures are applied, such as treating GIT and chest infections with proper antibiotics. A trial of gamma-globulin replacement therapy may be warranted in patients with recurrent infections if prophylactic antibiotics fail to diminish the number of infections. This may be administered intravenously or subcutaneously. Gamma-globulin replacement, however, does not result in appearance of the IgA at the mucous membranes of the respiratory tract. Again, a potential complication of intravenous immunoglobulin therapy in patients with anti-IgA antibodies is anaphylaxis. To minimize the occurrence of anaphylactic reactions, immunoglobulin preparations with the lowest content of IgA should be utilized whenever possible. Some patients may have an associated selective IgG2 deficiency and family history of common variable immune deficiency can be present in others. ***(Correct Answer: d).***

5) A 41-year-old female presents with pain all over her body. She says that her spine, limbs, and chest have been painful for the past 2 weeks, which are somewhat alleviated by paracetamol tablets. One year ago, she underwent lumpectomy with localized radiotherapy for stage II right-sided breast cancer. She's been doing well until 2 weeks ago when she started to feel out of power with aches. She is on no regular medication for the time being. Examination of the breasts shows only the scar of previous surgery and there are no palpable axillary lymph nodes. Serum alkaline phosphatase is raised. There are hypercalcemia and leuko-erythroblastic blood picture. Many liver target lesions are found on abdominal ultrasonographic examination. Per cutaneous biopsy of the hepatic lesions reveals breast secondary tumors, which are estrogen receptor-negative and demonstrate very low level of HER2. Both lung fields are bombarded with cannon balls. She is willing to receive any treatment and is desperate for help. She keeps saying, "I don't want to die."

Choose 2 stems, as part of this woman's treatment plan?

a) Cranial irradiation

b) Bone marrow transplantation

c) Partial hepatic resection

d) Zoledronic acid

e) Breast radiotherapy

f) Radical mastectomy

g) Trastuzumab

h) Chemotherapy

i) Pulmonary irradiation

j) Anastrozole

Objective: Review the management of metastatic breast cancer.

Around 10% of women with breast cancer have metastatic disease at the time of diagnosis; the majority develops such dissemination after being treated for localized or locally advanced breast cancer. This woman obviously has wide-spread disease and she requests treatment.

Treatment should be frankly discussed with all such patients addressing that cure is very unlikely, complete remission from chemotherapy is uncommon, and that treatment carries considerable toxicity for getting a marginal survival benefit. However, those who respond well are usually young, have limited disease (oligometastatic disease), and with excellent pre-treatment functional status.

Our patient's tumor is estrogen receptor-negative; therefore, endocrinal manipulation (Tamoxifen, Anastrozole…etc.) is virtually useless. Trastuzumab (Herceptin®) has been shown to be effective in tumors which express high levels of HER2; those with very low (as in our patient) or no HER2 don't benefit from this mode of therapy. Localized radiotherapy can be delivered for locally painful skeletal areas, but irradiation of lung metastatic tumors is not applicable. The question did not give a clue to intracranial secondary tumors; as a result, stem "a" can be cancelled. Surgical resection of localized breast secondary may be applied in selected cases, but our patient has no such a thing; radical mastectomy is used in stage I or II disease. She has bone marrow involvement as evident by the blood film's result; this may respond to chemotherapy (there is no place for bone marrow transplantation). Bisphosphonates have been shown to reduce skeletal events in metastatic (lytic, and probably blastic as well) breast cancer. ***(Correct Answer: d, h)***.

6) A 39-year-old plumber presents with anorexia, vomiting, weight loss, and upper abdominal discomfort for the past 3 months. His father and older sister died of familial polyposis coli. He had a positive genetic testing for familial polyposis coli gene mutation and underwent total removal of the colon at the age of 22 years. He has been doing well since then, and he enjoys an independent life, and lives with his wife in an apartment. He denies doing drugs, and he neither smokes nor drinks alcohol. His GP sees him at 1-year intervals and he undergoes blood tests at each visit. Reviewing his notes reveal no abnormal tests. He takes monthly vitamin injections and jogs every morning for 30 minutes in the local park.

What would you do for this man?

a) Refer him to a psychiatrist

b) Repeat genetic testing for a new mutation

c) Upper GIT endoscopy

d) Blood carcinoembryonic antigen (CEA)

e) Brain CT scan

Objective: Review the associations/complications of familial polyposis coli.

The commonest malignancy that occurs in familial polyposis coli (FPC) patients who underwent prophylactic colectomy at the appropriate age is duodenal cancer (especially the peri-ampullary one). This man with anorexia, weight loss, and vomiting must have his upper GIT scoped. Nothing in the question points towards Turcot's syndrome variant of FPC; therefore, brain CT scan is not indicated. Blood testing for CEA is usually done as part of post-operative management of colorectal cancer patients to detect metastasis or residual disease; it has no place in our patient. Assuming that his complaints are psychogenic in origin is not appropriate; a psychiatry referral will be counterproductive besides delaying the diagnosis. There is no new mutation in this man. ***(Correct Answer: c).***

7) A 67-year-old retired secretary presents with a mass in the right jaw angle that has been enlarging for 2 months. She has well-controlled hypertension on Lisinopril and her left knee osteoarthritis pain responds well to indomethacin. She has dry eyes and mouth for 10 years for which takes artificial lubrication in the form of methylcellulose eye drops and artificial saliva spray respectively. Examination reveals blood pressure of 135/85 mmHg, no raised JVP, normal chest, and benign abdomen. Her breasts and axillae are unremarkable. There is a mass that fits the right parotid region and is non-tender, 3 x 5 cm in maximum diameter, and no overlying skin changes. Neck lymph glands are non-palpable. Testing for ANA, anti-Ro and anti-La are positive.

What is the likely cause of the mass?

a) Primary Sjörgren's syndrome

b) Parotid pleomorphic adenoma

c) Jaw osteoma

d) Development of lymphoma

e) Breast secondary

Objective: Review the hematologic complications Sjögren's syndrome.

This woman has primary Sjögren's syndrome, as evidenced by her sicca symptoms and the positive serum antibody profile. This syndrome is a well-known cause of bilateral parotid enlargement. However, such rapid unilateral enlargement is always suspicious. The occurrence of rapid parotid enlargement in Sjögren patients should always prompt the physician search for the development of lymphoma (which occurs in 6% of cases).
Parotid (mass) biopsy should be done to define its histology. Parotid gland biopsy is not part of the work-up of Sjögren's diagnosis; instead, minor salivary glands biopsy is taken. Malignant lymphoproliferative disorders occur with a higher frequency in patients with Sjögren's syndrome. The spectrum of malignant lymphoproliferation extends from an increased frequency of monoclonal gammopathy, free light chains, and mixed monoclonal cryoglobulins (type II mixed cryoglobulinemia), to non-Hodgkin's lymphoma (NHL) or MALToma. The risk of NHL is up to 44-times higher than in the normal population. The life-time risk to an individual with primary Sjögren's syndrome is 4 to 10%. ***(Correct Answer: d)***.

8) A 63-year-old man, who was diagnosed with advanced gastric cancer, comes to see you because of his agonizing abdominal pains. He says that he takes his 60 mg controlled-release morphine twice daily but during the last few days, he has been experiencing severe deep gnawing pains in the upper abdomen and back. These have prevented him from sleeping, eating, and watching TV. He took paracetamol tablets but no improvement was noticed. He is extremely anxious and is desperate for your help.

What is the most appropriate thing to do?

a) Increase the controlled-release morphine to 120 mg twice per day

b) Add codeine 30 mg three times daily

c) Add Amitriptyline 50 mg at night

d) Ask him to take immediate-release morphine 20 mg when this pain occurs

e) Refer for celiac axis neurolysis

Objective: Review the management of pain in cancer patients.

This man has many breakthrough pains. The best approach is to keep him on his daily regimen and ask him to ingest 20 mg of an immediate-release (IR) preparation (and that is 1/6th of the daily controlled-release form) as a rescue measure, whenever these pains occur. Adding Amitriptyline will not address these severe breakthrough pains, but is useful as an adjunct in the long-term. Increasing the daily doses of the controlled-release (CR) morphine will not improve these breakthrough pains. Celiac axis neurolysis is used selected cases when the pain is refractory to potent opioids. IR morphine has an onset of action within 20 minutes that lasts for about 4 hours, while the CR form onset of action is 4 hours and its effect can continue for 12 hours. ***(Correct Answer: d).***

9) A 47-year-old woman seeks medical advice because of fear of breast cancer. She says that she read an article in the internet about breast cancer screening and she requests one. Her periods started at the age of 12 years and her cycles are regular at 28-day intervals. For the last few months, her cycles have been getting shorter with less blood. She has 3 children; all of them were breastfed. No family history of breast or ovarian cancer is present. She uses an intrauterine device for contraception. She has no hypertension or diabetes. BMI is 23 Kg/m^2.

Which one of the following is the correct statement regarding this woman's breast cancer screening?

a) Breast MRI is done now and at 5-year intervals if normal

b) Breast self-examination annually

c) Physician-conducted breast examination every 3 years

d) Annual mammography starting at the age of 50 years

e) BRCA1 and BRCA2 mutation analysis

Objective: Review cancer "screening".

This woman has an average risk for breast cancer; she has no personal or family history of breast or ovarian cancer, her cycles started at appropriate age, she breastfed her babies, and she is about to enter menopause. The best screening plan for this woman is monthly breast self-examination, 6-monthly physician-conducted breast examination, and annual mammography started when she becomes 50 years. BRCA1 and BRCA2 mutations should be suspected whenever there is strong family history of early onset or bilateral breast cancer (or ovarian cancer). ***(Correct Answer: d).***

10) A 63-year-old man presents for his annual check-up. He is an ex-smoker and does not drink nor does drugs. He exercises every morning by running in the local neighborhood for 20 minutes. He uses daily alpha methyldopa to control his hypertension and Finasteride for his prostatism. Examination is unremarkable. Chest X-ray shows a rounded irregular mass in the left upper lung zone, 1 x 2 cm in maximum diameter. He undergoes bronchoscopy with biopsy of the lesion. The histopathological diagnosis is squamous cell lung cancer. Chest CT scan does not reveal hilar or mediastinal lymph node enlargement as well as no pleural effusion. Hb 13 g/dl, serum potassium 3.9 mEq/L, blood urea 38 mg/dl, and serum calcium 10 mg/dl. Bone scanning is normal.

What is the best treatment option?

a) Neoadjuvant chemotherapy.

b) Chemotherapy.

c) Surgical removal.

d) External beam irradiation.

e) Hospice care.

Objective: Review the treatment modalities of non-small lung cancer according to its stage.

This man's bronchogenic carcinoma was detected incidentally. He has no symptoms. The overall clinical picture is that of T1N0M0 cancer; early non-small cell lung cancers are best treated with surgical excision if there are no contraindications. Small cell lung cancer is not treated by surgery. Neo-adjuvant (before surgical removal) chemotherapy in non-small lung cancer can used in stage IIIa of the disease (i.e., locally advanced cancer). Deep X-ray therapy in this man can be used as a palliative measure when there is pulmonary collapse, severe breathlessness, or massive hemoptysis. ***(Correct Answer: c)***.

11) A 72-year-old man is brought to the Emergency Room comatose. "My grandfather has metastatic prostate cancer and takes daily analgesics only," his grandson says. Examination shows Glasgow Coma Scale of 4/15, rapid shallow breathing, blood pressure 90/40 mmHg, bilateral coarse chest crackles, and severe global wasting.

Which one of the following is true with respect to this man's presentation?

a) Do brain CT scan

b) Echocardiography is worthy

c) Serum calcium, blood urea, and electrolytes can be useful

d) Invasive hemodynamic studies are indicated when replacing fluids

e) Give only O2 therapy

Objective: Review the management of terminally ill cancer patients.

This case highlights certain ethical issues which physicians can encounter when treating patients with advanced cancer. The "analgesics only" that he uses should make you conclude that his cancer is beyond any treatment option. His current presentation might reflect combination of chest infection, heart failure, metabolic derangement…etc. The question is "do we need to manage him aggressively?" What are the chances of survival? His general condition before coma is already poor and this coma may be the terminal event. After careful explanation of what he is going through to his family, application of general supportive measures (oxygen, fluids, analgesics...etc.) is all that is required until he passes peacefully. An ethical dilemma; not all people accept this idea and many may expect you to do everything to keep their patient alive! ***(Correct Answer: e).***

12) A 33-year-old female from Malaysia visits the United Kingdom. One week after arrival, she develops left-sided chest pain that turns out to be pulmonary infarction. There is right lower limb deep venous thrombosis. She has no chronic diseases. Her mother and older brother display a history of venous thrombosis. She has been taking daily oral contraceptive pills for 1 year. She has 2 healthy daughters; there is a history of 1 miscarriage. Her blood tests are unremarkable. PT and aPTT are within their normal reference range.

What else you will do for this woman for the time being?

a) Factor V Leiden mutation

b) Anti-phospholipid antibodies

c) No need for testing

d) Prothrombin gene mutation

e) Testing for anti-thrombin III deficiency

Objective: Know the appropriate "timing" for starting a work-up for suspected hereditary hyper-coagulable state.

This woman, who takes oral contraceptive pills, has developed leg's deep venous thrombosis (DVT) and pulmonary embolism after a long-haul travel (i.e., her event is provoked and there are 2 transient risk factors for this DVT). However, her family history is strong for a hereditary cause; she most likely has an underlying hereditary hyper-coagulable state that has come into light after the addition of these 2 transient risk factors. The most appropriate step for the time being is acute anti-coagulation. It is quite possible that this woman has a deficiency of factor C, factor S, or anti-thrombin III. It is not advisable to start a work-up for hereditary hyper-coagulable states at the time of presentation or after the start of anti-coagulation, as most of the tests will be affected by the event itself or by the anti-coagulation therapy.

This patient should be anti-coagulated for 3-6 months. At least 2 weeks should pass after the discontinuation of warfarin to start testing for factor C, factor S, or anti-thrombin III deficiencies. This case highlights a common clinical issue, where many physicians make inconsistent decisions whether to investigate first before starting anticoagulation or to investigate while patients are being anti-coagulated; both are wrong.

Factor V Leiden and prothrombin G20210A mutations are not found in native Asian populations. Her clinical picture is not suggestive of anti-phospholipid syndrome. ***(Correct Answer: c)***.

13) A 38-year-old woman visits the physician's office. She has just finished a 6-month course of warfarin for left leg deep venous thrombosis and pulmonary infarction. To avoid getting pregnant, she was using oral contraceptive pills for the past 2 years but she had stopped ingesting them when her deep venous thrombosis developed. She asks if she can resume her pills again. Few weeks after stopping warfarin, you do some tests and you find that she is heterozygous for the prothrombin gene G20210A mutation.

What would you tell her?

a) You can safely resume the pills, and no precaution is needed

b) Take daily low dose aspirin with the pills

c) Continue warfarin for another 3 months and then we reassess

d) Take the pills with low intensity warfarinization

e) Avoid taking the pills, and there is no need for further anticoagulation

Objective: Review the management of the first episode of "provoked" deep venous thrombosis.

This woman has developed her first, yet "provoked", episode of deep venous thrombosis with resultant pulmonary embolism; provoked (i.e., there is a transient risk factor) events carry a much lower risk of recurrence than the unprovoked episodes. Proper anticoagulation for 6 months would suffice in this woman, and all patients should be educated about the features of recurrence of thromboembolism and to consult their physician accordingly. Although this woman has a heterozygous mutation in the prothrombin G20210A gene, which is a risk factor for the very first "unprovoked" episode of venous thromboembolism, its presence does not confer an increased risk of recurrence. However, the addition of oral contraceptive pills would increase the latter risk and she should avoid using the contraceptive pills in the future; she should use an alternative way of contraception, instead. Although the use of low-intensity warfarin (with a target INR of 1.5 to 2) protects against recurrent venous thrombosis events without a significant increase in bleeding risk, this patient's relatively low annual recurrence risk in the absence of provocative risk factors would argue against the continued use of even low-intensity warfarin. ***(Correct Answer: e).***

14) You attend an international symposium about cancer screening. The symposium addresses the usefulness of various screening methods and their role in decreasing the mortality figure of each screened cancer.

Which one of the following screening methods decreases the mortality rate of the target cancer?

a) Annual chest X-ray for lung cancer in smokers

b) Six-monthly pelvic examination in hysterectomized women for ovarian cancer

c) Monthly self-conducted examination in pre-menopausal women for breast cancer

d) Biennial fecal occult blood in stool for colorectal cancer

e) Pap smear every 5 years for cervical cancer in sexually inactive women

Objective: Review cancer screening and recognize which methods reduce the risk of death in the target cancer.

Screening for colorectal cancer can identify premalignant lesions and detect asymptomatic early-stage malignancy. Doing fecal occult blood testing (FOBT) every 2 years has been shown by many randomized trials to decrease the mortality rate of colorectal cancer when compared with control subjects. FOBT screening has some disadvantages: it is not a good test for the detection of polyps, which usually do not bleed; and the sensitivity for advanced adenomas is substantially less than that for cancers and, therefore, it is necessary to work-up many false-positive results. ***(Correct Answer: d).***

15) A 24-year-old female has been found to have low platelets count on a routine pre-employment medical examination. She denies chronic illnesses and she does not take any medication. There is no similar family history. Her examination is unremarkable. She is single and is sexually inactive. Her blood tests show:

Hemoglobin:	14.1 g/dl
WBC:	6,300/mm^3
Platelets:	49000/mm^3
Blood film:	decreased number of platelets but there are no immature or abnormal cells.

What would you do?

a) Bone marrow study

b) Repeat platelets count at monthly intervals

c) Arrange for splenectomy

d) Start prednisolone

e) Do blood and urinary screen for drug abuse

Objective: Recognize the timing of starting medical therapy for immune thrombocytopenia.

This healthy-looking young woman has isolated thrombocytopenia. The differential diagnosis list includes hypo-proliferative states and peripheral consumption/destruction. The overall clinical picture does not provide a single clue to the underlying cause. The best approach would be monitoring her platelets count every month. The most likely diagnosis in this woman is immune thrombocytopenia; a presumptive diagnosis of idiopathic thrombocytopenic purpura is made when the history (e.g., lack of ingestion of a drug that can cause thrombocytopenia), physical examination, complete blood count, and examination of the peripheral blood smear do not suggest other etiologies for the isolated thrombocytopenia. Bone marrow studies are not routinely required to solidify the diagnosis of idiopathic thrombocytopenic purpura and the other stems are not justified. When the platelets count falls below 30,000/mm^3 in the absence of bleeding, medical treatment (prednisolone) is required.

It should be noted that surgical bleeding due solely to a reduction in the number of platelets does not generally occur until the platelet count is <50,000/mm^3, and clinical or spontaneous bleeding does not occur until the platelet count is <10,000 to 20,000/mm^3. ***(Correct Answer: b).***

Chapter 6

Rheumatology and Disease of Bones
10 Questions

This page was left intentionally blank

1) A 41-year-old housewife was diagnosed with rheumatoid arthritis 3 years ago. She has been taking daily sulfasalazine and weekly methotrexate since then. Today, she visits you with a concern about her eyes. She says that her eyes burn and she feels as if there is sand in them. Her hands are painful on movements and there is mild ulnar deviation of the fingers. Rheumatoid factor titer is positive, and there is normochromic normocytic anemia, raised ESR, and thrombocytosis.

What is the best treatment for her ocular complaints?

1) Give methotrexate daily
2) Daily oral prednisolone
3) Monthly gold injections
4) Methylcellulose eye drops
5) Topical dexamethasone eye drops

Objective: Recognize that the treatment of Sjögren's syndrome is largely symptomatic.

She has developed secondary Sjögren's syndrome (dry eyes, and possibly dry mouth). The best treatment of the ocular dryness is artificial eye drops (such as methylcellulose or polyvinyl alcohol), and artificial saliva sprays can be used for the oral dryness; note that the treatment of Sjögren's syndrome is largely symptomatic. Treating the underlying disease in secondary cases definitely needs attention but this will not alleviate the sicca symptoms. This woman absolutely needs modification of her current therapy to bring the disease into remission and control, but the question has addressed the treatment of the ocular complaints only. ***(Correct Answer: d).***

2) A 42-year-old factory worker presents with hand pain and stiffness for the last 5 months. He says that his hands are sore and are painful when he uses them and that they are especially stiff upon awaking in the morning. He has low back pain and his GP told him that it is a form of mechanical one. He denies neck pain, chest pain, bowel symptoms, and urinary complaints. He drinks a little whisky every night but does not do drugs. His older brother has a rigidity problem affecting his entire spine and has been receiving a treatment for it for the past 10 years. The patient has mild intermittent asthma, which is responsive to inhalers. He takes an injection of vitamin B_{12} every month as he thinks that this is a healthy habit. Examination reveals swelling of the distal interphalangeal joints of both hands, painful active and passive movements, and limited spinal movements with tenderness of the lower back. The large proximal joints are normal-looking. ESR is 60 mm/hour.

What does the man have?

1) Ankylosing spodylitis
2) Reiter's syndrome
3) Rheumatoid arthritis
4) Adult Stills disease
5) Psoriatic arthropathy

Objective: Review the mode of presentation of psoriatic arthropathy.

Although the question did not mention any skin lesion or nail manifestation, the combination of spinal complaints, *distal* interphalangeal joint swelling (and hand soreness) with raised ESR and positive family history of ankylosing spondylitis should alert you about the possibility of psoriatic arthritis. Familial clustering in seronegative spondarthritides is common and is not necessarily of the same disease. Many features overlap (e.g., skin, eye, enthesopathy…etc.) and some cases may not have a clear-cut diagnosis. Rheumatoid factor and ANA are negative. The first-line medications are NSAIDs; if no response occurs or the disease is severe and polyarticular, disease modifying anti-rheumatic drugs should be started. Methotrexate, cyclosporine, and retinoic acid derivatives have been shown to improve the skin and articular manifestations; however, none of them has been shown to retard or at least prevent joint damage. ***(Correct Answer: e).***

3) A 72-year-old man visits the physician's office. The man says that he has been diagnosed with Paget's disease of the bone 2 years ago, after having pain in his mid-spine and enlarged D_{10} vertebra. The back pain is responsive to mefenamic acid and he is not that bothered by that pain. He is house-bound and spends most of his time in bed reading. He denies headache or hearing problem. The patient' s legs do not seem to be bowed. His blood pressure is well controlled with Irbisartan and his blood lipids are responding to atorvastatin. Blood counts and renal function tests are normal. Serum calcium is 11.8 mg/dl. Serum alkaline phosphatase is twice its upper normal limit.

What is the *correct* statement?

1) He needs spinal fixation surgery

2) Prednisolone is better to be given

3) No treatment is needed

4) Bone scanning should be done

5) Start oral tiludronate

Objective: Review the indications of starting medical therapy for Paget's disease of the bone.

The usual indications of treatment with specific anti-Pagetic medications (bisphosphonates and calcitonin) in symptomatic patients are neurological complications (nerve compression, basilar invagination…etc.), headache due to skull involvement, and prominent pain syndromes not responsive to simple analgesics, as well as immobilization hypercalcemia (as in our patient). Bowed legs will not correct themselves and deafness is unlikely to improve, while pain due to secondary osteoarthritis may or may not improve upon receiving such therapy. The indications of treatment of *asymptomatic* patients are less clear. However, treatment can be started when any one of the following is present:

1. Extensive skull involvement.
2. Disease involving the tibia or femur in which progression is likely.
3. Serum alkaline phosphates that is at least 4 times its upper normal limit (i.e. indicating moderately active disease) in patients with bone involvement at sites liable for severe complications (such as near joints, weight-bearing joint, vertebral body…etc.).

Our patient is minimally symptomatic but has immobilization hypercalcemia. He should receive medical therapy. ***(Correct Answer: e).***

4) A 50-year-old woman comes to see you with a problem of having fatigue most of the time. She also says that her neck, shoulders, axial spine, and limbs are painful. She denies any joint swelling but she admits to having recurrent colicky abdominal pain, alternating diarrhea and constipation, and urinary frequency. She insists that she was healthy and had no chronic illnesses, and that she enjoyed an independent life. Her sleep is fragmented and her mood is low. She neither smokes nor drinks alcohol. She is single and has no children. Her GP ascribed these features to cessation of menstruation. However, you consider fibromyalgia. From the following list, choose 2 options that are *inconsistent* with your preliminary diagnosis?

a) ESR 50 mm/hour

b) TSH 3.0 mU/L

c) Hemoglobin 13 g/dl

d) Serum alkaline phosphatase 70 iu/L

e) Negative serum ANA

f) Normal chest X-ray

g) Serum ALT 15 iu/L

h) Eosinophil count 2%

i) Distended colonic bowel loops on abdominal ultrasound

j) Platelets 90 x 10^9/L

Objective: review the work-up of fibromyalgia.

The combination of symptomatology, normal lab investigation, and multiple tender points can secure the diagnosis of fibromyalgia. The blood tests should be limited to routine ones, as well as TSH and muscle enzymes, which all should be normal. More extensive testing is not indicated and may even give confusing results, e.g., ANA can be found in normal healthy individuals, and unless the clinical picture is suggestive of SLE, ordering this test will be counterproductive. Stem "i" can be seen in irritable bowel syndrome (which this woman also manifests).

Raised ESR and thrombocytopenia are never normal accompaniment of fibromyalgia and their presence should cast a strong doubt on the diagnosis. The list of differential diagnoses of fibromyalgia is long and includes hypothyroidism, depression, polymyalgia rheumatica, connective tissue disorders (especially rheumatoid arthritis, SLE, and Sjögren's syndrome), inflammatory myositis and metabolic myopathy. Fibromyalgia is a chronic pain disorder that is difficult to treat. The cornerstone of the treatment is reassurance and careful explanation of the disease combined with a low dose of a tricyclic agent at night, graded aerobic exercises, and cognitive behavioral therap. ***(Correct Answer: a, j).***

5) A 57-year-old woman is referred to you for further evaluation. Rheumatoid arthritis was diagnosed 1 month ago, after the patient had had hand pain and stiffness for 3 months. You examine the patient and order some blood tests. Finally, you disagree with your colleague's diagnosis.

What have you detected?

a) Negative serum rheumatoid factor.

b) Lumbar spine pain and limitation of movements.

c) Lower limbs skin ulceration.

d) Finger gangrene.

e) Bland urinary sediment.

Objective: Recognize features that stand against the diagnosis of rheumatoid arthritis.

Rheumatoid arthritis is rheumatoid factor-negative in 30% of cases; the absence of rheumatoid factor does not refute the diagnosis. Rheumatoid arthritis does not produce nephritis and the urinary sediment is expected to be bland. Rheumatoid vasculitis can produce skin ulceration and gangrene and even gangrene of the fingers or toes. Apart from attacking the cervical spine, rheumatoid arthritis usually spares the lower spine; features of prominent lumbar spondylitis and/or sacroiliitis should cast a strong doubt on the diagnosis. ***(Correct Answer: b)***.

6) A 43-year-old man has been experiencing pain and stiffness in both hands and feet over the past several months. He also admits to having paresthesia in both lateral palms. Has stage I hypertension, which is well-controlled with hydrochlorothiazide. Blood tests reveal Hb 9.1 g/dl, blood urea 39 mg/dl, and serum uric acid 8.1 mg/dl. 12-lead ECG shows 3:1 AV block. This is his left upper limb:

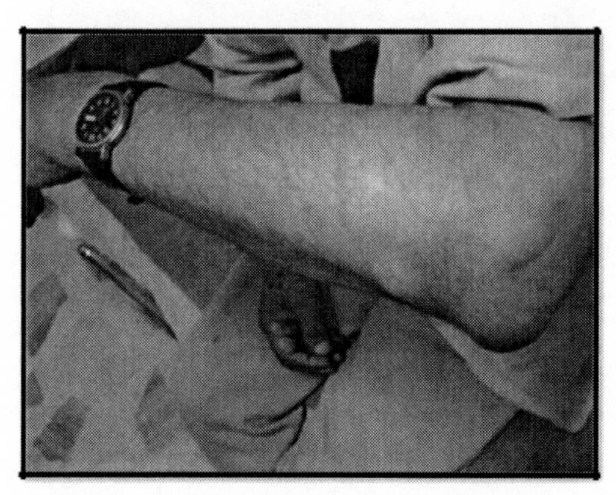

Which one of the following with respect to this man's diagnosis is *correct*?

a) Serum rheumatoid factor is most likely negative.

b) His spleen could be enlarged.

c) Excessive watering of the eyes is an expected finding.

d) The forearm lesions are not tophi.

e) The hands' sensory disturbance results from cervical radiculopathy.

Objective: Review the clinical features of rheumatoid arthritis.

The mildly raised serum uric acid in this patient (normally, up to 7.0 mg/dl) is common with long-term use of diuretics. This, together with the forearm lesions may misdirect you to think of gouty arthritis. Every question must have a clue; the type II second-degree AV block cannot be explained by gouty arthritis (or its medical therapy).

The constellation of high-grade AV block, "bilateral carpal tunnel syndrome", and mild anemia, pain and stiffness in hands and feet is consistent with rheumatoid arthritis (RA). Accordingly, this nodular RA must be sero-positive. A palpable spleen in RA patient is encountered in Felty's syndrome; this syndrome develops after several years (an average of 10 years) with deforming but inactive RA (our patient's clinical course spans several months only). The commonest ocular manifestation of RA is Sjögren's syndrome (with dry eyes). Bilateral C_6 (lateral palm dermatome) radiculiopathy has no solid explanation in this patient, and carpal tunnel syndrome is an expected complication of wrist arthritis. The lesions in the image are rheumatoid nodules. ***(Correct Answer: d).***

7) A 63-year-old man has been referred by his GP to your office as a difficult-to-manage case of chronic sinusitis. You find left sided proptosis. He admits to having bloody nasal discharge occasionally. He is unable to dorsiflex his left heel and there is wasting of the right hypothenar eminence. He has fixed pulmonary opacities on the chest plain films and 2 pulmonary nodules. These are his blood tests:

Hemoglobin:	9.1 g/dl
WBCs:	16 x 10^9/L
Neutrophils:	12.7x 10^9/L
Eosinophils:	0.3 x 10^9/L
Platelets:	550 x 10^9/L
ESR:	90 mm/hour
Blood urea:	79 mg/dl
Urine:	proteinuria, hematuria with RBCs casts
Serum potassium:	5.1 mEq/L
Serum sodium:	139 mEq/L
ANA:	negative
Rheumatoid factor:	negative
TSH:	2.4 mU/L

What diagnosis this man has?

a) SLE

b) Churg-Strauss syndrome

c) Polyarteritis nodosa

d) Wegener's granulomatosis

e) Disseminated tuberculosis

Objective: Recognize the characteristic features of systemic vasculitides.

Superficially, you may have concluded that this man has some sort of vasculitis. Analyze his findings. He has chronic sinusitis, bloody nasal discharge, and unilateral proptosis. In addition, he has mononeuritis multiplex. The chest X-ray film shows infiltrates and nodules. There is anemia, neutrophilic leukocytosis (with *normal* eosinophil count), thrombocytosis, and raised ESR; these are markers of active disease. The renal function is impaired and he has active urinary sediment. Each type of vasculitis has a clue.

The presence of proptosis and lung nodules would favor Wegener's granulomatosis over other types of systemic vasculitides. The normal eosinophil count and no history of asthma would cancel Churg-Strauss syndrome. The negative ANA would make SLE highly unlikely; the sinus disease, lung nodules, and proptosis are incompatible as well. Polyarteritis nodosa usually results in renal infarcts rather than frank glomerulonephritis. The whole picture is not suggestive of tuberculous infection. ***(Correct Answer: d)***.

8) A 57-year-old man visits the physician's office for a scheduled follow-up. He complains of exertional breathlessness and you find bi-basal crackles. This is his plain film of the hands.

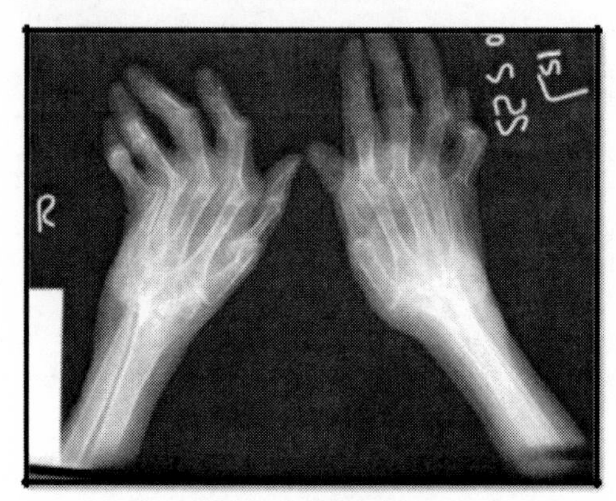

All of the following about this man's disease are incorrect, *except*:

a) Interstitial pulmonary fibrosis is unlikely

b) The development of heart failure should cast a doubt on the diagnosis

c) A blowing early diastolic murmur down the left lower sternal border suggests infective endocarditis

d) Palisading granulomas never found in the sclera

e) The axillary lymph nodes may be palpable

Objective: Recognize the extra-articular manifestations of rheumatoid arthritis.

The plain film shows changes consistent with long-standing rheumatoid arthritis. Rheumatoid arthritis is a multi-systemic disease. Lymph nodes draining actively inflamed joints may become secondarily enlarged; the development of lymphoma or methotrexate-induced pseudo-lymphoma is a differential diagnosis. The palisading granulomas of rheumatoid nodules are found in the lungs, pericardium, pleural, skin, and sclera. The respiratory tract, from the larynx down to the lung parenchyma may be involved, and the development of interstitial pulmonary fibrosis is no exemption.

Myocardial dysfunction, pericarditis, conduction defects, aortitis, and aortic reflux all may occur. ***(Correct Answer: e).***

9) A 71-year-old man visits the Emergency Room. He has severe pain in the right metatarsophalangeal joint, which is also red and extremely tender. He has chronic stable angina, hypertension, hypercholesterolemia, and osteoarthritis of the left knee. This is the 4th visit to the district's Emergency Department within 6 months. You do the following investigations:

Hemoglobin	15.2 g/dl
WBC	12,300/mm^3
Blood urea	55 mg/dl
Serum creatinine	1.8 mg/dl
Serum uric acid	9.1 mg/dl
Random plasma glucose	143 mg/dl
Serum calcium	10.1 mg/dl

Which one of the following is the best treatment option for this time?

a) Colchicine

b) Prednisolone

c) Intramuscular diclofenac

d) Anakinra

e) Allopurinol

Objective: recognize that pre-existent renal impairment influences the choice of acute gouty arthritis treatment.

There is mild renal impairment. Oral colchicine can cause severe GIT upset while the intravenous colchicine would be very toxic and may result in severe bone marrow suppression in this man. Oral or parenteral NSAIDs may well worsen his mild renal function. Allopurinol is indicated in patients with recurrent acute gouty arthritis and hyperuricemia, but should only be started after the acute attack has subsided. Glucocorticoids are the safest agents in elderly patients who have renal impairment.

Anakinra is a recombinant IL-1 receptor antagonist and has been shown to be effective in the treatment of acute gouty arthritis; however, this treatment is still investigational and is not recommended. ***(Correct Answer: b)***.

10) A 30-year-old typewriter visits the physician's office complaining of severe intermittent pain in his fingers, especially upon cold exposure. He also says that his both feet are painful, mainly upon walking. The GP has prescribed some medications for sciatica. He drinks occasionally and smokes 2 packets of cigarettes every day. You find splinter hemorrhages, blue right 2nd and 3rd toes, and black left big toe. The leg pulses are very feeble. He denies doing drugs, and his family history shows that his older sister has a cardiac problem. Serial blood cultures are negative and echocardiography is reported as normal. You do aortography and find normally patent aorta, corkscrew collaterals, and sudden vascular cutoffs at both wrists and ankles.

What is the best treatment?

a) Combined intravenous antibiotics

b) Intravenous thrombolysis

c) Monthly high dose intravenous methylprednisolone

d) Daily cyclophosphamide

e) Smoke cessation

Objective: Review the mode of presentation and treatment of Buerger's disease.

Buerger's disease is a form of vasculopathy of small and medium-sized arteries and veins, particularly at the distal extremities (upper and lower). The proximal vessels are usually normal and the internal organs are rarely involved. Most victims are young men who smoke heavily; the use of tobacco is the *sine qua non* of disease *initiation and progression.* Superficial thrombophlebitis can occur as a very early disease manifestation (even before digital ischemia is clinically evident) and may parallel disease activity. Reynaud's phenomenon occurs in approximately 40% of cases (as in our patient). A cardiac source of embolization must be ruled out by echocardiography. Angiography of *both* arms and legs should be performed, even in patients who present with clinical involvement of only one extremity because of the high prevalence of disease in multiple limbs.

The angiographic findings of Buerger's disease, such as segmental occlusion and collateralization around areas of occlusion (corkscrew collaterals), while suggestive, are not pathognomonic, since they may be identical to findings in patients with other causes of small vessel occlusive disease, e.g., scleroderma, CREST syndrome, mixed connective tissue disease, and other cause of vasculitis.

This man should quit smoking immediately, completely, and for good. About 94% of patients who quit smoking avoid amputation compared with 43% who continue to smoke. Trans-dermal nicotine patches or nicotine chewing gum may keep the disease active, but bupropion can be used as a smoking cessation aid. Surgical revascularization is usually not recommended due to the diffuse segmental involvement and distal nature of the disease. ***(Correct Answer: e)***.

This page was left intentionally blank

Chapter 7

Endocrinology and Metabolic Medicine
15 Questions

This page was left intentionally blank

1) A 34-year-old woman comes to the physician's office because of her thyroid problem. She was diagnosed with hypothyroidism, 1 year ago, because of Hashimoto's thyroiditis. Her only current medication is L-thyroxin, 100 μg per day. She denies weight gain, constipation, aches and pains, or menstrual irregularities. Her mother has primary atrophic hypothyroidism and her father has diabetes and hypertension. She is active at work and she is planning to direct a new TV series. Examination shows pulse rate of 70 beats per minute and blood pressure of 110/85 mm Hg. Serum T_4 is 80 nmol/L (normal 60-145 nmol/L) and serum TSH is 15 mU/L (normal 0.5-5.0 mU/L).

What would you do next?

a) Decrease the daily L-thyroxin dose

b) Add growth hormone injections

c) Increase the daily L-thyroxin dose

d) Keep the same daily dose of L-thyroxin

e) Add cortisone

Objective: Review the treatment and follow-up of hypothyroidism.

Although the patient is relatively symptomatically well and her daily dosage of L-thyroxin is stable in the past 8 months, her lab results are disappointing. The objective of treating primary hypothyroidism is to bring the serum TSH down to its normal reference range; this patient's serum TSH is far from this. Simply, increasing the daily dose (say to 125 μg per day) with serial follow-ups is needed to see which daily dose keeps this serum TSH within its reference range (keeping her with the same daily dose is not acceptable). The patient is not over-treated (no suppressed TSH), in order to decrease the daily L-thyroxin dose. The hypothyroidism is not secondary to pituitary problem (e.g., low ACTH resulting in adrenocortical failure); therefore, there is no need for glucocorticoid replacement therapy. For the same reason, growth hormone replacement has no place. When starting treatment for primary hypothyroidism, serum T_4 normalizes first and the serum TSH lags behind for a few weeks and even months; this should be kept in mind in the follow-up visits. ***(Correct Answer: c)***.

2) A 54-year-old pharmacist visits the physician's office because of lethargy. He says that he cannot cope with his drugstore work and this lethargy has been increasing over the last 9 months. He has lost interest in sex and does not go to his favorite local diner at weekends, as he used to do. He reports nausea, loss of appetite, and constipation. He feels dizzy when he stands suddenly from a low chair. His past medical history shows successfully treated post-primary pulmonary tuberculosis, 1 year ago. Examination reveals a downcast face, tanning, thin physique, and postural hypotension. His blood urea is 55 mg/dl, serum potassium is 5.8 mEq/L, and his blood sugar is 65 mg/dl.

What is the best next step?

a) Complete blood count

b) Serum calcium

c) Morning serum ACTH and cortisol

d) Serum growth hormone during sleep

e) Renal biopsy

Objective: Recognize the causes of adrenal failure.

This man's old chest tuberculosis might well have had involved the adrenals to produce Addison's disease, which is responsible for this man's current non-specific presentation. Addison's disease can mildly elevate blood urea and may produce hyperkalemia, hypercalcemia, hypoglycemia, anemia, and eosinophilia. Early morning serum cortisol is usually high in normal people; low serum cortisol levels with high serum ACTH should call for further evaluation of Addison's disease. Short ACTH stimulation test should be performed in all patients in whom the diagnosis is being considered. Growth hormone deficient patients are usually plumb and may have fatigue, but the deficient hormone does not produce skin tanning or the blood results mentioned in the question. Renal failure patients may present solely with malaise, but there is earthy-colored face rather than generalized tanning, and the degree of renal impairment is much deranged. Stems "a" and "b" are used in the overall management plan of Addison's disease, but they are not diagnostic by their own. ***(Correct Answer: c).***

3) A 45-year-old man is referred to by his GP as a newly diagnosed case of acromegaly. The patient is hypertensive and has bilateral carpal tunnel syndrome. His head seems to be large. He has been subjected to many lab tests. Blood tests show normal blood counts and ESR, hyperglycemia, glycosuria, and normal renal function. Insulin-like growth factor-I (IGF-I) level is in the lower part of its reference range. The GP has already arranged measurement of serum growth hormone and brain MRI.

What would you do next?

a) Start pegvisomant

b) Discuss the matter with the neurosurgeon

c) Take the opinion of the anesthetist for fitness under general anesthesia

d) He has no acromegaly

e) Do insulin tolerance test

Objective: Review the diagnosis of growth hormone excess.

Hyperglycemia, bilateral carpal tunnel syndrome, and hypertension all can occur in acromegaly. The presence of the above features in someone with a "large" head has alerted the GP about the possibility of acromegaly and he has already arranged a diagnostic plan. Most diagnostic guidelines start with measurement of the IGF-I; if this is normal, acromegaly can be excluded and no need for further testing. However, if IGF-I is elevated, one can confidently proceed with measurement of serum growth home during oral glucose tolerance test. If the latter test shows inappropriate suppression (adequate suppression of serum growth hormone is inconsistent with acromegaly), brain MRI is the next step to image the pituitary. If the pituitary is normal, chest and abdomen CT scanning is done with measurement of the serum GHRH level. This man demonstrated a low normal level of serum IGF-I; this means that he has no acromegaly. Pegvisomant is an analogue of human growth hormone; it selectively binds to growth hormone receptors, blocking the binding of endogenous GH, leading to decreased serum concentrations of insulin-like growth factor-I (IGF-I) and other GH-responsive proteins. It is used to treat growth hormone deficient patients. ***(Correct Answer: d).***

4) A 69-year-old retired office manager has a 3-year history of body aches and pains, paresthesia in both hands, constipation, and depressed mood. She is not responding to fluoxetine and her GP is planning to use imipramine. She takes herbal remedies. She neither smokes nor drinks alcohol. Her past surgical history is unremarkable. One of her sisters has schizophrenia. She is obese, has pale puffy face, blood pressure 160/95 mmHg, and wasting of both thenar areas. Blood tests show Hb 8 g/dl, MCV 108 fl, blood urea 30 mg/dl, ALT 13 iu/L, creatine phosphokinase 340 u/L, blood sugar 92 mg/dl, and serum cholesterol 290 mg/dl.

What would you do next?

a) Combine fluoxetine and imipramine

b) Bone marrow study

c) Brain MRI

d) 24-hour urinary free cortisol

e) Serum TSH and T_4

Objective: Review the lab findings in hypothyroidism.

The woman's history is suggestive of a major depressive illness in someone with a family history of psychosis. However, her lab tests are abnormal; macrocytic anemia, hypercholesterolemia, and raised serum CPK enzyme. The bilateral carpal tunnel syndrome (note the wasted thenar eminences), obesity, and hypertension would suggest hypothyroidism. Choosing TSH and T_4 would seem reasonable. Serum creatine phosphokinase is not raised in Cushing's syndrome and carpal tunnel syndrome is not an integral part of that syndrome. Primary hypothyroidism is much more common that the secondary one; therefore, brain MRI is not indicated, as there are no clues to a possible intracranial pathology (head trauma, brain surgery, bitemporal field defect…etc.). Anemia in hypothyroidism can be macrocytic (associated pernicious anemia), normochromic normocytic (of chronic diseases), and even hypochromic microcytic (from iron deficiency due to menorrhagia in middle-aged women); bone marrow study is not indicated in general. Replacing the deficient thyroid hormone will improve her depressive symptoms; antidepressants alone have no substantial effect.

Her sister's illness might be accidental and is not related to our patient's disease, or might represent one of the neuropsychiatric manifestations of hypothyroidism (vague personality changes, depression, psychosis-myxoedema madness, confusional state, or frank dementia). ***(Correct Answer: e)***.

5) A 26-year-old clerk with a 9-year history of type I diabetes mellitus presents with frequent hypoglycemic episodes. He denies any change in his daily work, insulin dosage/frequency, or missing a major meal. He spends most of his daily working hours in his office and does not exercise. His symptoms are palpitations, sweating, tremor, and anxiety, which are relieved by candy ingestion. These mainly develop in the morning hours of work. During the last 4 days, his fasting plasma glucose was as follows: 77, 56, 68, and 47 mg/dl. His blood pressure is 100/70 mmHg, supine, that becomes 70/50 mmHg upon standing. He has undergone few blood tests, which have revealed Hb 11.3 g/dl, ESR 12 mm/hour, HbA1c 6.8%, blood urea 53 mg/dl, serum potassium 5.9 mEq/L, serum total bilirubin 0.6 mg/dl, and ALT 11 iu/L.

Which investigation would you choose to discover the cause of these hypoglycemic episodes?

a) Plasma insulin.

b) Plasma C-peptide.

c) Serum ACTH.

d) Autonomic function testing.

e) Gastric emptying scintigraphy.

Objective: Review the causes of hypoglycemia in diabetics.

The commonest causes of hypoglycemia are errors in insulin dosage/administration, missing a major meal, and an unexpected increase in physical activity. These are not present in our patient who seems to have a sedentary life style and well-controlled blood sugar (note the HbA1c <7%). The presence of these fasting hypos, together with postural hypotension and hyperkalemia with slightly raised blood urea should always prompt the physician search for Addison's disease development. Morning plasma ACTH will be high in Addison's disease and proceeding to short ACTH stimulation test is the correct next step. Don't forget that type I diabetes is an autoimmune disease that might be associated with other autoimmune diseases, such as pernicious anemia, thyroid diseases, Addison's disease…etc.

Dysautonomia in longstanding diabetes does not cause hypoglycemia, but gastroparesis may contribute to fluctuating blood sugar control; this man has no gastric symptoms and gastric emptying studies are not indicated, neither autonomic function testing (the orthrostasis might misdirect you to choose this). ***(Correct Answer: c)***.

6) A 54-year-old man visits the physician's office because of feeling weak all the time. He says that he sleeps well at night for a minimum of 7 hours. He works as an accountant at an iron factory. He has stage I hypertension, which is treated by hydrochlorothiazide. He denies fever, sweating, headache, or orthopnea. He had a pituitary acidophil adenoma that was treated by cranial irradiation, 15 years ago. Examination reveals striking pallor, smooth skin, and loss of axillary and pubic hair.

How would you approach this patient?

a) Serum prolactin

b) Brain MRI

c) Dynamic pituitary hormonal testing

d) ECG

e) Chest X-ray

Objective: Review the diagnosis of pituitary failure.

This man, who has a history of "acromegaly", was treated with cranial radiotherapy. Whether this was the primary modality of treatment or had been applied after surgical removal of the pituitary acidophil adenoma, the overall presentation is that of pituitary failure. The best step for the time being is to do dynamic (or static) pituitary hormonal assessment. Serum prolactin alone is useless. Brain imaging is done next to visualize the sellar area; the gland could have a normal or small size; recurrent tumor is unlikely to be seen after these 15 years. The question did not mention any gross features of acromegaly (large hands with spade-like fingers, prognathism, kyphosis…etc.); the clue is the acidophil adenoma. All patients with pituitary masses, whether treated or not, need regular follow-up. In all endocrinal dysfunction, the biochemical diagnosis should precede the imaging study. ***(Correct Answer: c).***

7) A 2-year-old child presents with recurrent pancreatitis. He had no perinatal trauma or infections. His mother had not taken any drug or medication during pregnancy. His notes show normal milestones. Examination reveals tubero-eruptive xanthomas but no xanthelasmas. Your colleague suggests doing fundoscopy for him.
What do expect to see upon doing fundoscopy?

a) Retinitis pigmentosa

b) Cherry red spot

c) Central retinal artery occlusion

d) Lipemia retinalis

e) Optic atrophy

Objective: Review the complications of hypertriglyceridemia.

The presence of recurrent pancreatitis in a child with these types of xanthomas is highly suggestive of severe hypertriglyceridemia due to APO CII or lipoprotein lipase deficiencies. This hyperlipidemia has a characteristic retinal sign; lipemia retinalis. Severe hypertriglyceridemia is a risk factor for central retinal vein, not artery, thrombosis. Neither retinitis pigmentosa nor cherry red spots are part of this disease. The treatment is usually difficult with dietary management and multiple lipid-lowering drugs. ***(Correct Answer: d).***

8) A 41-year-old alcoholic man is admitted to The Emergency Room because of repeated vomiting. He is emaciated and hypotensive. Resuscitation is started with fluids, carbohydrates, and vitamins and he improved a little. However, after few hours he started to develop cardiac dysrhythmia, myalgia with raised creatine phosphokinase, falling hemoglobin, and indirect hyperbilirubinemia, and finally he becomes confused.

What is the reason for this deterioration?

a) Central potine myelinolysis

b) Hypokalemic paralysis

c) Nutritional recovery syndrome

d) Alcoholic ketoacidosis

e) Alcohol withdrawal

Objective: Review the causes of hypophosphatemia.

The combination of cardiac dysrhythmia, CNS dysfunction, hemolysis, and muscle weakness (raised CPK) can be explained by nutritional recovery (or re-feeding) syndrome. Actually, this syndrome is the commonest cause of severe hypophosphatemia. This usually occurs in chronic alcoholics who have malnutrition and total body deficit of phosphate, and who are then given high amounts of carbohydrates. Central pontine myelinolysis presents with quadriparesis upon rapid correction of severe hyponatremia. Agitation, tremulousness, sweating…etc. are features of alcohol withdrawal. Nothing in the question is suggestive of alcoholic ketoacidosis. Hypokalemic paralysis cannot be explained by confusion and hemolysis. ***(Correct Answer: c)***.

9) A 27-year-old female presents with a 4-week history of anxiety and easy irritability. She underwent thyroid removal, 1 year ago, because of Graves' disease. She has been doing well since then and she takes no medication. You find fine and fast postural hand tremor and rapid irregular heart rate. Her blood tests reveal serum total bilirubin of 1.4 mg/dl, serum free T_4 of 2.2 ng/dl (normal, 0.9-2.4 ng/dl), and serum calcium of 11 mg/dl. Brain MRI is unremarkable. This is her neck:

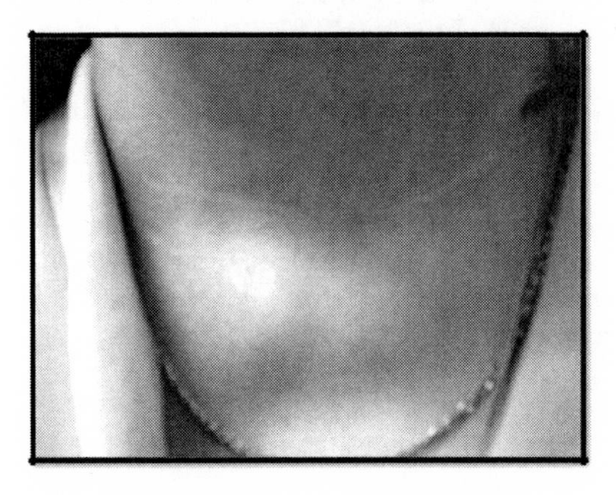

What is the reason behind her current complaints?

a) Struma ovarii

b) Co-existent pheochromocytoma

c) Surreptitious thyroid hormone ingestion

d) Alcohol withdrawal

e) Recurrent Graves' disease

Objective: Review the follow-up of thyroidectomized patients.

The image above shows a "collar" scar of previous thyroidectomy; the bulging masses at the lower neck were actually lipomas! This woman had improved substantially after removing the bulk of her thyroid gland; however, she has recently developed symptoms and signs that suggest a hyper-adrenergic state.

Patients who are not hypothyroid soon after surgery need to be monitored for possible future hypothyroidism or recurrent hyperthyroidism for the rest of their lives. Recurrent hyperthyroidism occurs in 2% of patients. This patient most likely has developed a recurrence.

Hyperthyroidism may result in many "non-specific" lab abnormalities; liver functions are commonly deranged and serum calcium may be raised mildly. The normal serum level of the free T_4 hormone means she has *T_3 toxicosis*; note that serum TSH was not given in the question! The phrasing of the question may misdirect you to choose struma ovarii or co-existent pheochromocytoma. Approximately 5 to 15% of women with struma ovarii have hyperthyroidism due to a thyroid adenoma in the struma. Although surreptitious thyroid hormone ingestion is a possibility in this woman, recurrence of the thyroid gland hyper-functioning is much more common clinically, as there are no clues in the question towards this option (e.g., very low serum thyroglobulin). ***(Correct Answer: e).***

10) Because of impotence, a 31-year-old man visits the physician's office. His blood tests are:

Serum prolactin	23 ng/ml (normal, 2-15 ng/ml)
Serum FSH	0.1 iu/L (normal, 0.9-15 iu/L)
Serum LH	0.3 iu/L (normal, 1.3-13 iu/L)

Which one of the following statements about this man's illness is *correct*?

a) No need to proceed with dynamic testing for other pituitary hormones

b) Serum testosterone is most likely low normal

c) Microprolactinoma is the most likely cause

d) Imaging of the brain is indicated

e) Old mumps orchitis could be relevant

Objective: Review causes of central hypogonadism.

A very abbreviated scenario has been given and the data were provided sparingly. This is a very common form of questioning in written examinations. The occurrence of impotence with very low FSH and LH means hypogonadotrophic hypogonadism, resulting in secondary lowering of serum testosterone. This is together with the raised serum prolactin may misdirect you towards microprolactinoma. Again, each question must have a clue. The *very mildly* elevated serum prolactin with profound suppression of serum FSH and LH should prompt the physician search for mass lesions in/around the pituitary (which might have resulted in disconnection hyperprolactinemia). This man needs dynamic pituitary hormone assessment and imaging of the brain. Testicular damage (old orchitis, tuberculosis, trauma, surgical removal,…etc.) would result in secondary elevation in serum LH and FSH. ***(Correct Answer: d).***

11) A 63-year-old man presents with poor exercise tolerance. This is his face:

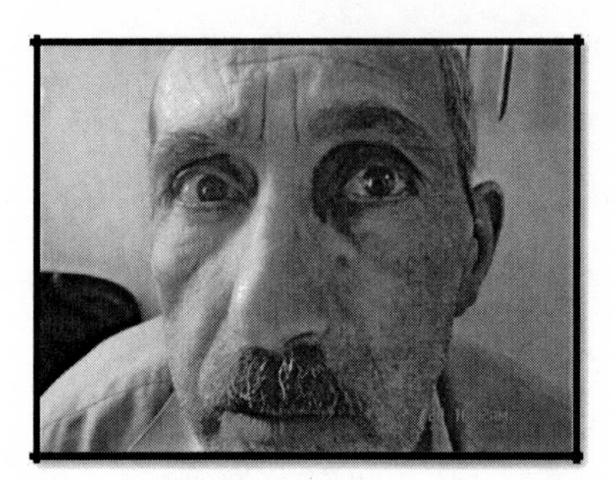

The patient denies orbital or head trauma. Brain/orbital CT scan is unremarkable. He says that his facial appearance has resulted from a minor eye rubbing. You find peripheral sensori-motor polyneuropathy. Urine is shows ++ protein. Transthoracic echocardiography reveals marked left ventricular hypertrophy. 12-lead ECG shows wide-spread low voltage QRS complexes.

What is the likely diagnosis this man has?

a) Wegener's granulamtosis

b) Polyarteritis nodosa

c) Acute myeloid leukemia

d) Sarcoidosis

e) Primary AL amyloidosis

Objective: Review the characteristic clinical features of primary AL amyloidosis.

Signs of skin involvement in systemic amyloidosis include waxy thickening, easy bruising (ecchymosis), and subcutaneous nodules or plaques. This purpura, which is characteristically elicited in a periorbital distribution (raccoon eyes) by a valsalva maneuver or minor trauma, is present in only a minority of patients, but is highly characteristic of AL amyloidosis.

The combination of low voltage ECG complexes with marked LVH on echocardiography is also characteristic of cardiac amyloidosis. ***(Correct Answer: e).***

12) A 17-year-old girl is referred by her GP as a difficult-to-manage case of epilepsy. Her blood tests show TSH 2.4 mU/L (normal, 0.5-5), ACTH 79 pg/ml (normal, 9-52), growth hormone (after a provocation test) 20 ng/ml (normal, >7), fasting blood glucose 95 mg/dl, stool fat (on a 100-g fat diet) 3 g/d (normal, <5 g/d). The GP states that the girl's nails are abnormal, discolored, and fragile. This is her non-contrast brain CT scanning:

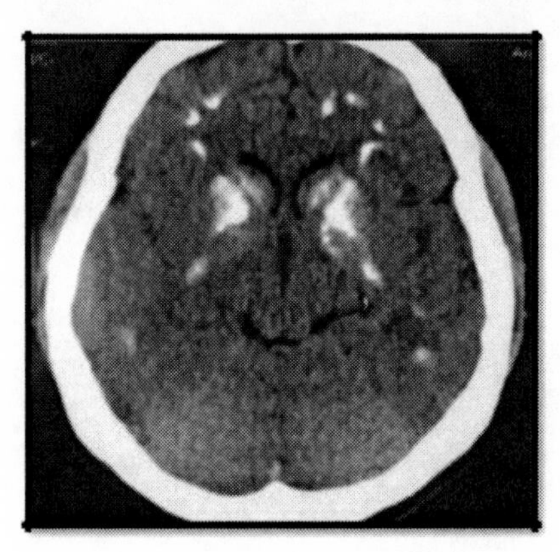

What does the girl have?

a) Polyglandular autoimmune syndrome type I

b) Multiple endocrine neoplasia type I

c) McCune Albright Syndrome

d) Celiac disease

e) Mitochondrial encephalopathy, lactic acidosis, and stroke-like episodes (MELAS)

Objective: Review the components of polyglandular autoimmune syndromes.

The constellation of epilepsy, basal ganglia calcification, adrenal failure (raised serum ACTH), and candidiasis (note the nails description) in an adolescent girl is highly suggestive of polyglandular autoimmune syndrome type I. Serum PTH and calcium should be measured next. Hypoparathyroidism or chronic mucocutaneous candidiasis is usually the first manifestation, characteristically appearing during childhood or early adolescence, always by the early twenties.

The hypoparathyroidism may, or may not, occur in association with anti-parathyroid gland antibodies that are directed against the calcium-sensing receptor. Serum growth hormone *normally* rises after a provocation stimulus; this is together with the raised serum ACTH may misdirect the candidate towards choosing hyperpituitarism. ***(Correct Answer: a)***.

13) A 49-year-ol man visits the physician's office. He has diabetes and ischemic heart disease. The patient is HIV-positive and has been receiving efavirenz/emtricitabine/tenovovir (Atripla®) since 2 years. HIV plasma viral load is <400 copies/ml. Lipid profile shows LDL-cholesterol of 170 mg/dl. You prescribe a statin for the latter observation.

Which one of the following statins you have prescribed?

a) Simvastatin

b) Atorvastatin

c) Rosuvastatin

d) Fluvastatin

e) Pravastatin

Objective: Recognize that rosuvastatin is the safest statin in efavirenz-containing anti-HIV medications.

HIV infection is associated with highly atherogenic lipid profile, which eventually contributes to stroke and coronary events in this population. The mechanism is multi-factorial. Efavirenz, which is a protease inhibitor, reduces the blood concentration of most statins (60% for simvastatin, 40% for pravastatin, and 34% for atorvastatin). This will ultimately produce under-treatment of dyslipidemia. Rosuvastatin has no known interactions with efavirenz and is the most useful statin in this man. ***(Correct Answer: c).***

14) A 60-year-old woman undergoes dual-energy X-ray absortiometry testing. She is white, post-menopausal, and has a BMI of 22 Kg/m^2. She has never experienced a bone fracture before. These are her results:

Spine:	T score= -2.7, Z score= -1.3
Hip:	T score= -1.3, Z score= -0.8

According to the WHO criteria for defining bone mass, what diagnosis this woman has?

a) Normal bone mass

b) Osteopenia

c) Severe osteopenia

d) Osteoporosis

e) Established osteoporosis

Objective: Review the WHO criteria for defining bone mass.

The World Health Organization (WHO) established a classification of bone mineral density (BMD) according the standard deviation (SD) difference between a patient's BMD and that of a young-adult reference population (T-score). A BMD T-score that is 2.5 SD or more below the young-adult mean BMD is defined as *osteoporosis*, provided that other causes of low BMD have been ruled out (such as osteomalacia). A T-score that is 1 to 2.5 SD below the young-adult mean is termed *osteopenia* (or low bone mass). *Normal* bone density is defined as a value within one standard deviation of the mean value in the young adult reference population. *Established* osteoporosis is defined as a bone density in the osteoporotic range with one or more fragility fractures. Note that the WHO categorization system does not include "*severe*" osteopenia and that this system uses the T-score only. The Z-score is a comparison of the patient's BMD to an age-matched population. A Z-score of -2.0 or lower is considered below the expected range for age. Thus, the presence of Z-score values more than two standard deviations below the mean should prompt careful scrutiny for coexisting problems (e.g., glucocorticoid therapy or alcoholism) that can contribute to osteoporosis.

This woman's spinal T-score is -2.7; this is osteoporosis. The hip T-score is -1.3; this defines osteopenia. Since osteoporosis is present in the spine, this diagnosis takes precedence over the osteopenia that is present in the hip of this woman. ***(Correct Answer: d)***.

15) A 29-year-old woman has been referred to your office by her gynecologist for further evaluation. She is primigravida, at her 24th week of gestation. Yesterday, she underwent a screening glucose tolerance test using a 50-gram glucose load. At the end of the 1st hour, her plasma glucose turned out to be 151 mg/dl. Her mother has type II diabetes. Her past notes do not reveal any abnormal plasma glucose testing.

What would you do next?

a) Start dietary measures to control her diabetes

b) Reassure her; she has no gestational diabetes and no need for further testing

c) Start metformin

d) Measure urinary ketones

e) Proceed to 3-hour glucose tolerance test using a 100-gram glucose load

Objective: Review the screening/diagnostic approach of gestational diabetes.

Pregnant women with *any* of the following appear to be at increased risk of developing gestational diabetes: a family history of diabetes, especially in first degree relatives; pre-pregnancy weight ≥ 110% of ideal body weight or body mass index over 30 Kg/m^2 or significant weight gain in early adulthood, between pregnancies, or in early pregnancy; age greater than 25 years; previous delivery of a baby greater than 4.1 Kg; personal history of abnormal glucose tolerance; member of an ethnic group with higher than the background rate of type 2 diabetes (in most populations, the background rate is approximately 2%); previous unexplained perinatal loss or birth of a malformed child; maternal birth weight greater than 4.1 Kg or less 2.7 Kg; glycosuria at the first prenatal visit; polycystic ovary syndrome; current use of glucocorticoids; and essential hypertension or pregnancy-related hypertension.

On the other hand, a pregnant woman must have *all* of the following characteristics to be labeled as having "low risk" of developing gestational diabetes: age less than 25 years; normal weight before pregnancy (BMI less than 25 Kg/m^2); member of an ethnic group with a low prevalence of gestational diabetes (i.e., patient is not Latino, African-American, Native American, South or East Asian, Pacific Islander); no first degree relative with diabetes mellitus; no history of abnormal glucose tolerance; and no history of poor obstetric outcome.

Screening should be performed at the 24th to 28th weeks of gestation. However, screening should be done as early as the first prenatal visit if there is a high degree of suspicion that the pregnant woman has unrecognized pre-gestational diabetes. Screening is done with the 1-hour glucose tolerance test using a 50-gram glucose load. If the 1-hour plasma glucose is $\geq$130 mg/dl, the diagnostic 3-hour glucose tolerance test should be done next using a 100-gram glucose load. If two or more of the following serum glucose values are met or exceeded, the patient has gestational diabetes:

1. Fasting serum glucose concentration >95 mg/dl (5.3 mmol/L).

2. One-hour serum glucose concentration >180 mg/dl (10 mmol/L).

3. Two-hour serum glucose concentration >155 mg/dl (8.6 mmol/L).

4. Three-hour serum glucose concentration >140 mg/dl (7.8 mmol/L).

***(Correct Answer: e)*.**

Chapter 8

Infectious Diseases and Genitourinary Medicine

10 Questions

This page was left intentionally blank

1) A 45-year-old homosexual airplane pilot presents with fever and skin rash for the last 1 week. He has oral candidiasis and palpable neck lymph glands. The pharynx is congested and there is mild neck stiffness. He admits to having unprotected sex with men during his trips and occasionally doing drugs. His current daily medications are amlodipine for hypertension and simvastatin for hypercholesterolemia. He denies vomiting, weight loss, or breathlessness. After careful counseling, he agreed to undergo HIV testing. Plasma HIV RNA viral load is 210,000 copies/ml and the CD4+ count is 260 cells/mm^3.

All of the following statements with respect to his current diagnosis are wrong, *except*:

a) A prolonged illness portends rapid progression to AIDS

b) Oral ulceration is against the diagnosis

c) Atypical lymphocytes in blood indicate co-existent EBV infection

d) Neck stiffness is due to cryptococcal meningitis

e) The patient is relatively non-infectious

Objective; Review the mode of presentation of acute HIV seroconversion illness and the risk of progression to AIDS.

The patient displays a high-risk behavior for HIV infection; although his sex with men is protected, but he is still at risk of acquiring the infection. Acute HIV seroconversion illness usually occurs 2-4 weeks, post-exposure; up to several months may elapse before the appearance of this syndrome in some patients. The HIV viral load in plasma is typically high and the viremia makes the patient highly infectious to others. Oral ulcerations are seen in 20% of cases, and the meningitic process is due to the virus itself. Factors that may indicate rapid progression to AIDS are oral thrush, low CD4+ count, prolonged illness, and presence of opportunistic infections at the time of diagnosis. Epstein-Barr viral mononucleosis ranks the first on the list of the differential diagnoses; heterophile antibodies are uncommonly positive in HIV seroconversion illness and will pose a diagnostic dilemma. ***(Correct Answer: a).***

2) A 19-year-old college student from Dublin presents with fever and skin rash for 5 days. He returned from Thailand, 1 week ago. He spent 2 months there as part of students exchange program. He also reports retro-orbital pain, severe bone and muscle pains, and nausea and vomiting. He does not remember any insect bite. He declined vaccination before travelling, 2 months ago. He took no prophylactic medications for infections and had no sex there. As far as he knows, he has not come in contact with sick people. He takes daily sodium valproate for migraine prophylaxis and he denies doing drugs. His sister has common variable immune deficiency syndrome. Examination reveals fully conscious patient, temperature 39.1 °C, diffuse maculopapular rash, conjunctival injection, pharyngeal redness, and palpable liver edge. Bloods show leukopenia, thrombocytopenia, and mildly raised serum AST.

What does the man have developed?

a) Yellow fever

b) Classic dengue fever

c) Crimean-Congo hemorrhagic fever

d) Dengue hemorrhagic fever

e) Acute HIV seroconversion illness

Objective: Differentiate between dengue fever and dengue hemorrhagic fever.

This patient travelled for the first time to an area that is endemic in dengue. It is the commonest mosquito-borne infection world-wide (over 100 million dengue viral infections occur annually throughout the world). The patient did not recall a mosquito bite; however, the bite might be minor and easily forgotten by the patient. Absence of such bite does not refute the diagnosis of dengue. After exposure to the virus, the incubation period is 3-14 days; dengue can be excluded when the traveler develops fever after 2 weeks of his/her return. Our patient developed the typical features of classical dengue fever. About 55% of infections after the first exposure are either asymptomatic or minimally symptomatic. Dengue hemorrhagic fever (note the absence of hemorrhagic manifestations in this man) needs a *past* exposure to the wild virus to occur.

To date, there is no licensed vaccine to protect against dengue. Stems "a" and "c" are not seen in Thailand. Stem "e" can be safely excluded by the absence of sexual history or other high-risk behaviors (e.g., drug abuse). ***(Correct Answer: b)***.

3) A 43-year-old homosexual bartender is brought to the Emergency Department. He is confused and irritable and has right-sided weakness. His past records show CD4+ cell count of $87/mm^3$ and high plasma HIV RNA load. You phone his GP, who says that the patient stopped taking anti-HIV medicines, 1 year ago, because of GIT upset and currently he is on no medication apart from certain herbals. Brain CT scan reveals 2 deep hemispheric rounded masses, surrounding edema, and mass effect.

What would you do next?

a) Start highly active anti-retroviral therapy

b) CSF examination

c) Stereotactic brain biopsy

d) Anti-toxoplasma therapy

e) Whole brain irradiation

Objective: Review the diagnostic approach and treatment of space-occupying lesions in HIV-infected patients.

The top differential diagnoses in this man, who has low CD4+ cell count and no antiretroviral therapy or prophylactic medications, are cerebral toxoplasmosis and primary CNS lymphoma. The scenario is not biased to either of these. Cerebral toxoplasma encephalitis is more common than primary CNS lymphoma and when toxoplasma serology is positive, the diagnosis is more consolidated (although definite diagnosis is by brain biopsy). All such patients should receive a trial of anti-toxoplasma treatment and observed closely for 1-2 weeks. If the presentation is atypical (say, with visual field defect only), toxoplasma serology is negative (we don't know our patient's one), or there is no improvement after anti-toxoplasma therapeutic trial in terms of clinical and radiological parameters (after 7-10 days), stereotactic brain biopsy should be contemplated. Around 20% of primary CNS lymphoma patients have involvement of the ocular vitreous; aspiration of vitreous can secure the diagnosis without undergoing brain biopsy. The cerebral mass effect should call against doing CSF examination (which is likely to be of little value in this patient). Sure, this patient needs HAART to improve the immune status, but this is not the urgent/required step to treat his brain condition. ***(Correct Answer: d).***

4) A 37-year-old homosexual man visits the Emergency Department. He has been experiencing shortness of breath and low-grade fever for the past 2 weeks. He is allergic to sulfa drugs. He work-up is shown below:

Hemoglobin	9.1 g/dl
CD4+ cell count	126/mm^3
Blood urea	62 mg/dl
Urine	++++ protein
Serum LDH	870 iu/L
ALT	13 iu/L
Serum total bilirubin	0.9 mg/dl
Serum albumin	2.9 mg/dl
DLCO	61% predicted value
Chest film	perihilar ground glass opacities
Induced sputum	Unremarkable

All of the following statements about this man's current presentation are wrong, *except*:

a) Renal and urinary abnormalities reflect pre-renal failure

b) Chest X-ray is inconsistent with *P. carinii* pneumonia

c) Serum LDH portends a good prognosis

d) He should receive clindamycin, primaquine, and prednisolone

e) Highly active anti-retroviral therapy (HAART) should delayed for at least 2 months

Objective: Review drug allergies in HIV-infected patients and their alternative treatments.

Homosexuality is a risk factor for IHV infection. The question has not included a positive testing for HIV, but it has mentioned a low CD4+ count. This man has *P. carinii* (renamed as *Pneumocystis jiroveci*) pneumonia (PCP) as an AIDS-defining illness; the plain chest film is consistent with that diagnosis. A raised serum LDH in this infection portends a bad prognosis, as is the presence of hypoxemia (note the indirect evidence of hypoxemia by demonstrating a low DLCO). The prominent proteinuria and mild renal impairment should make you think of HIV nephropathy (glomerulosclerosis). The outlook of severe PCP has been shown to improve with the use of glucocorticoids. Because of sulfa allergy (i.e., co-trimoxazole is contraindicated), he should receive a combination of clindamycin and primaquine. HAART can be started. ***(Correct Answer: d).***

5) A 32-year-old TV reporter is planning to travel to Kenya the next week where he is going to stay for 4 weeks and then he will return back to Edinburgh. He visits your office seeking an advice about malaria prevention. He has no chronic diseases and denies drug allergies.

What would you do?

a) Advise against going there; cancel the trip

b) Atovaquone/proguanil

c) Weekly intravenous quinine

d) Chloroquine

e) Double strength trimethoprim/sulfamethoxazole

Objective: Review malaria prophylaxis and the dilemma of drug resistance.

Malaria is an important cause of fever and serious illness in returned travelers. The relative risk of malaria is higher among returned travelers from Sub-Saharan Africa than those from Asia or the Americas. Sub-Saharan African countries are endemic with malaria strains that are resistant to chloroquine. Actually, chloroquine-resistant strains are found all over the world; chloroquine as a prophylactic agent may be used in areas where its resistance is very low, e.g., the Caribbean, parts of China, Central America. The best options in this man are atovaquone/proguanil (Malarone®), doxycycline, or mefloquine; their choice depends on their cost, availability, and their adverse reactions. Prophylaxis should begin 1 week before reaching the malarious area and should continue for 4 weeks after leaving it. Trimethoprim/sulfamethoxazole has no significant anti-*Plasmodium* activity and, therefore, plays no role in the chemoprophylaxis of any form of malaria. Intravenous quinine is a very good choice in *treating* severe malaria. There is no reason to prevent this man from going to Kenya. Prevention efforts should be aimed at *all* forms of malaria. Most chemoprophylaxis regimens are designed to prevent primary attacks of malaria but they may not prevent the later relapses that can occur with *P. vivax* and *P. ovale*. ***(Correct Answer: b).***

6) You have been consulted by the oncology department to see this 43-year-old man, who has been receiving chemotherapy for his acute myeloid leukemia, M_2-sybtype. At the end of his 1st week of hospitalization, he developed a very painful vesicular rash on his belly. You diagnose *Herpes zoster* and you suggest isolation precautions.

Which one of the following isolation precautions you have suggested?

a) Negative pressure with airborne isolation

b) Contact isolation

c) Droplet isolation

d) Airborne and contact isolation

e) Standard isolation and there is no need for specific isolation precautions

Objective: Review isolation precautions and their indications.

Infection control is a discipline that applies epidemiologic and scientific principles and statistical analysis to the prevention or reduction in rates of nosocomial infections. *Herpes zoster* virus in patients with shingles can be transmitted to others via direct contact and airborne ways; combined contact and airborne isolation precautions are indicated in this patient. The other methods alone would not suffice. Standard precautions are recommended in the care of *all* hospitalized patients. ***(Correct Answer: d).***

7) A 61-year-old businessman from Aberdeen consults you about his intention to visit Brazil. He will stay there for 4 weeks and may visit the Amazon basin. He has surfed the internet and found that he may get certain infectious diseases if he goes there. He requests an advice about any preventive measures. He has prostatism, mild COPD, and well-controlled hypertension. You tell him that there are effective vaccines that can prevent some of these infections.

Which vaccine have you recommended?

a) Meningococcal, A and C

b) Oral typhoid

c) Cholera

d) Yellow fever

e) Influenza

Objective: Recognize the indications for vaccination according to the traveler's destination.

This man is planning to visit an area that is endemic with yellow fever. Yellow fever remains an endemic and epidemic disease problem of considerable magnitude, which occurs mainly in regions of Africa and South America. A highly effective live-attenuated vaccine (yellow fever 17D vaccine) was developed in 1936. Immunization is recommended for travelers to yellow fever endemic areas of Africa and South America, and for residents of those areas. Although he has COPD and he should receive annual influenza vaccine, the latter is not the most appropriate step for the time being. ***(Correct Answer: d)***.

8) A 60-year-old man has been found to a have temperature of 38.9 °C this morning. He has been receiving ABVD regimen for stage IIB Hodgkin's lymphoma, and this is the 4th cycle. Apart from fever, he does not report anything else. You examine him and found no lesions in the mouth, anus, and skin. The skin surrounding his central Hickman catheter shows no signs of inflammation. You take blood and urine cultures and start intravenous gentamycin and cefepime. His absolute neutrophil count (ANC) is 110/mm^3. The cultures return with no growth. After 3 days, the patient is still febrile and you notice redness and tenderness along the tunnel of Hickman catheter. You remove the catheter, culture its tip, and add intravenous vancomycin. After 1 week, the patient's temperature is 39.4 °C and the ANC is 134/mm^3. The patient's complaints have not changed and you have not noticed any new physical signs.

What would you do next?

a) Add intravenous amphotericin B to the current antibiotic regimen

b) Add intravenous levofloxacin to his antibiotic regimen

c) Continue the same antibiotic regimen for another 1 week

d) Stop the current antibiotics; add fluconazole

e) Stop the current antibiotics; Add quinupristin/dalfopristin

Objective: Recall the management of febrile neutropenia in cancer patients.

Fever in a neutropenic patient is defined as a single temperature of >38.3°C (101.3°F), or a sustained temperature >38°C (100.4°F) for more than one hour. The definition of neutropenia varies from institution to institution but is usually defined as an absolute neutrophil count <500 cells/mm^3 or <1,000 cells/mm^3 with a predicted nadir of <500 cells/mm^3.

The chemotherapy this man receives has resulted in profound and prolonged neutropenia. Prolonged neutropenia increases the risk of *Staphylococci*, Gram negatives, *Candida*, and *Aspergillus* infections. Cefepime, gentamycin, and vancomycin would cover *Staphylococci* and Gram negatives, including anaerobes. Vancomycin was added because of catheter tunnel infection; otherwise, it is not routinely added.

The persistence of high fever in spite of receiving this regimen should always prompt the physician think of invasive fungal infection and add amphotericin B to the current antibiotic regimen, even if no fungus has been isolated by smears and/or cultures. Simply stopping his antibiotic regimen is not justified, even if fever persists. Quinolones do not cover fungi, and quinupristin/dalfopristin alone is very defective in tackling febrile neutropenia with unknown microbial cause. ***(Correct Answer: a).***

9) A 23-year-old housekeeper presents with a 1-day history of suprapubic pain and painful micturition. General urine examination shows many pus cells and you diagnose acute urinary tract infection (UTI). She says that she developed a similar infection, 3 weeks ago, which responded favorably to trimethoprim/sulfamethoxazole tablets. You prescribe a 3-day course of the same antimicrobial. After 4 days, she comes back and says that her symptoms were alleviated in the first day only and then recurred afterwards.

What is the reason behind her worsening after an apparent improvement?

a) Re-infection with the same microbe

b) Re-infection with another microbe

c) Medications' non-compliance

d) Microbial resistance

e) Wrong diagnosis of UTI

Objective: You should be familiar with antimicrobial resistance of uropathogens in your community, and differentiate resistance from re-infection.

Recurrent urinary tract infections (UTIs) are common in women, result in considerable morbidity and expense, and can be vexing management problems for clinicians. However, there is no evidence that recurrent UTI leads to health problems such as hypertension or renal disease, in the absence of anatomic or functional abnormalities of the urinary tract. In this patient, documentation of re-infection with the same or another pathogen requires culture before and after finishing the treatment, and then at the second presentation. In addition, re-infection within 1 day after an initial improvement is highly unusual.

Most women with UTI respond to even very low doses of antibiotics, and non-compliance is not a major issue when treating UTI patients without physiological or anatomical defects of the urinary tract. Many uropathogens are resistant to trimethoprim/sulfamethoxazole, at least partially, in many communities. The prior trimethoprim/sulfamethoxazole exposure in this woman increases this possibility, and the improvement during the 1st day of treatment confirms that she has a UTI. The best approach would be choosing an alternative empirical antibiotic while pending urine culture. ***(Correct Answer: d.)***.

10) A 61-year-old woman has been transferred from a rural hospital to your tertiary center. She has progressive respiratory failure. Her transfer notes state that the patient was admitted to that hospital after developing ischemic stroke, 2 weeks ago. She was in the intensive care unit for 1 week. Chest X-ray at the day of transfer shows 2 pneumonic patches in the left lung field. One day before the transfer, her WBC was 21,000/mm^3 with shift to left. The referring hospital did tracheal aspirate for her, for Gram stain and culture, 5 days ago. The results came in yesterday showing *Acinetobacter*, that is resistant to many antibiotics. In addition, there are many sputum leukocytes, and sputum staining reveals Gram-negative coccobacilli.

What would you give at this time?

a) Ceftriaxone

b) Vancomycin

c) Imipenem

d) Ciprofloxacin

e) Spiramycin

Objective: Review the treatment of hospital-acquired pneumonia and recall the specific management of Acinetobacter.

This woman has developed hospital-acquired pneumonia with a multidrug resistant strain of *Acinetobacter*. Nosocomial transmission is responsible for the vast majority of *Acinetobacter* infections. Patients at risk are often critically ill with multiple co-morbidities, concurrent infections, and on prolonged courses of antibiotics. As a result, it may be difficult to distinguish between colonization and true infection. Pneumonia is the most common manifestation of nosocomial *Acinetobacter* infection, accounting for 6.9% of Gram-negative pneumonias in the ICU. Many guidelines acknowledge that a carbapenem (e.g., imipenem) *alone* is a reasonable choice, and that there is no evidence for improved outcome with combination therapy. Few aminoglycosides, such as amikacin, are also effective; gentamycin is not. The other options are ineffective. ***(Correct Answer c).***

This page was left intentionally blank

Chapter 9

Gastroenterology and Hepatology
15 Questions

This page was left intentionally blank

1) A 32-year-old primary school janitor comes for his scheduled check-up. He has been given a diagnosis of duodenal ulcer 4 weeks ago, after doing upper GIT endoscopy. The latter showed 2 x 3 cm duodenal bulb ulcer but no biopsy was taken. Treatment was started at that time with triple therapy consisting of lansoprazole, amoxicillin, and clarithromycin. Today, he says that he has the same initial symptoms of epigastric hunger pain that is relieved by food. He denies medication non-compliance. He does not smoke or drink alcohol. His mother died of esophageal cancer 2 years ago, at the age of 65 years. His older brother underwent gastric surgery for gastro-esophageal reflux disease.

How would you respond?

a) Repeat esophagogastroduodenoscopy and take biopsies

b) Barium meal

c) Barium-follow-through

d) Repeat the same treatment for another 4 weeks

e) Urea breath testing

Objective: Recognize the management of anti-H. pylori medication failure.

The 2 commonest causes of medical treatment failure of peptic ulcers are medications noncompliance and *H. pylori* resistance to antibiotics. Our patient demonstrates compliance with his treatment regimen; the second likelihood seems reasonable. To diagnose non-eradication of *H. pylori* in this patient, urea breath test can be done; if this turns out to be positive, it means that the organisms is still there and active, and the symptoms can be ascribed to treatment failure (whichever the reason was). Although the endoscopist did not take a biopsy from his duodenal ulcer, repeating the endoscopy is needed only in gastric ulcers (even if they are benign-looking) to document complete healing. Barium meal lacks sensitivity and specificity for peptic ulcers. The organism might well have resistance to anti-*H. pylori* medical treatment and continuing the same regimen would sound unwise. Barium-follow-through have no place in the diagnosis or follow-up of duodenal ulcers. *H. pylori* organisms are naturally resistant to several commonly used antibiotics, including vancomycin, trimethoprim, and sulfonamides. ***(Correct Answer: e)***.

2) A 34-year-old man is referred to you from the hepatology department for further management. The referral letter states that the patient is HBs-antigen positive. This was detected as part of medical insurance examination. He is an immigrant from Asia. He denies a previous history of jaundice. He practices sex with men and does intravenous drugs. He denies weight loss, itching, nausea, or malaise. He lives alone, drinks a unit of alcohol at night, and smokes a packet of cigarettes every day. Examination fails to show hepatomegaly, ascites, or jaundice. Blood tests show negative serum HBe antigen testing, elevated plasma HBV DNA, and persistent elevation in serum aminotransferases. HIV and hepatitis C viral tests are negative. Serum albumin and PT are normal.

What does the man have?

a) Hepatocellular carcinoma

b) Liver cirrhosis

c) Chronic hepatitis B infection with a pre-core mutant form

d) Chronic hepatitis C infection

e) HIV-induced peliosis hepatis

Objective: Review the serology of hepatitis B infections.

The patient is at risk of hepatitis B, hepatitis C, and HIV infections because of homosexuality and intravenous drug abuse. The latter 2 have been excluded by appropriate testing. There are no clues to the presence of hepatocellular carcinoma (liver mass, prominent elevation in alpha-fetoprotein…etc.) or cirrhosis (compensated or uncompensated). The question has addressed hepatitis B status of the patient. HBs antigen marks infection, and the elevated plasma HBV DNA level indicates high infectivity of the virus; the negative serum HBe antigen (a marker of infectivity) would normally cancel the latter possibility. The combination of chronic hepatitis B infection with elevated plasma viral DNA, negative HBe antigen, and persistent (or intermittent) elevation in liver aminotransferases (that cannot be ascribed to other causes) reflects an infection with an HBV-variant (usually a pre-core mutant or core-promoter variants). ***(Correct Answer: c).***

3) A 54-year-old woman presents with frequent defection of bulky stools. She admits to losing weight of 15 Kg over the past few months but she insists that she eats well and her appetite is good. She denies fever, blood in stool, or regurgitation. She has neither hypertension nor diabetes, and there is no family history of note. She underwent gastric surgery for perforated peptic ulcer, 2 years ago. Examination reveals pallor, global wasting, and normal blood pressure.

Which one of the following is inconsistent with bacterial overgrowth syndrome?

a) High serum folate

b) Low serum vitamin B12

c) Anemia with dimorphic blood picture

d) Watery diarrhea

e) Clubbing

Objective: Recall the clinical features of bacterial overgrowth syndrome.

Bacterial overgrowth (BOG) may complicate Billroth II gastrectomy. The bacteria consume the intra-luminal vitamin B_{12} and produce excess folate, which is then absorbed (its serum levels are therefore, high). The presentation may be that of watery diarrhea, rather than frank steatorrhea. Clubbing is not seen in BOG but may occur in cases complicating Crohn's disease. Non-invasive diagnosis can be secured with ^{14}C-glycocholate breath test. The cornerstone in the diagnosis is small bowel aspirate for bacterial culture. ***(Correct Answer: e).***

4) A 54-year-old man with alcoholic liver cirrhosis, Child-Pugh B, is brought to the Emergency Room. He is agitated and confused, and has prominent impairment of registration and recall. His daughter says that her father was reasonably well at home, until 2 days ago, when he gradually started to be drowsy and out of temper. "I also gave him his medications, so that he would not miss a dose," she added. He stopped ingesting alcohol, 1 year ago. He takes daily vitamins and tonics, oral lactulose, and propranolol. She does not recall her father having had fever, diarrhea, vomiting, bloody stool, or any change in his daily medicines over the past 2 weeks. Examination shows disorientation to time, place, and person, pallor, jaundice, global wasting, tense ascites, clubbing, and leukonychia. Blood sugar is 90 mg/dl, blood urea 50 mg/dl, and urine examination shows no abnormalities. Brain CT scan is unremarkable but chest X-ray shows small bilateral pleural effusions.

What else you will do?

a) Brain MRI

b) Serum bilirubin, AST, and ALT

c) Upper GIT endoscopy

d) PT and aPTT

e) Abdominal paracentesis

Objective: Recall the mode of presentation of spontaneous bacterial peritonitis.

This man presents with acute confusional state on the background of relatively compensated alcoholic cirrhosis. There must be a reason and that the reason must be identified and corrected rapidly. Such decompensation is usually due to diarrhea/constipation, development of renal failure (spontaneous or diuretic-induced), hypokalemia, GIT bleeding, infection (especially spontaneous bacterial peritonitis and UTI), high protein diet, trauma (including surgery), introduction of a centrally depressant agent, and large portosystemic shunting. Most of these causes can be excluded in our patient. Spontaneous bacterial peritonitis can simply present as decompensation and confusion, without any abdominal signs and symptoms, and should always be kept in mind in those with tense ascites. Ordering PT and aPTT, serum bilirubin, AST, and ALT will give an idea about the severity of liver dysfunction, but will not identify the cause of this sudden confusion.

His brain CT scan is unremarkable; therefore, ordering brain MRI is a waste of time (no added benefit; besides being difficult in confused patients). He has no features suggestive of upper or lower GIT bleeding and upper GIT endoscopy is not required. ***(Correct Answer: e)***.

5) A 21-year-old man presents with unintentional weight loss for 8 months. His stool is pale and bulky. You order some tests; the results are:

Fecal fat:	70 mmol (normal, <18)
Fecal occult blood:	Negative
Stool microscopy:	Undigested food particles, no parasite, blood or pus
Stool culture:	Negative
IgA anti-endomysium:	Negative
IgG anti-tissue transglutaminase:	Positive
MCV:	110 fL
Blood film:	Hypochromic RBCs
Blood urea:	34 mg/dl
Serum potassium:	3.8 mEq/L
ALT:	12 iu/L
Duodenal biopsy:	Subtotal villous atrophy

You tell him the diagnosis and you start treatment. He makes a favorable response in terms of weight gain and reducing stool volume. However, after 6 months, he returns back with the same original complaints.

a) What is the reason for his new visit?

b) Infection with a different organism.

c) Development of fistulae.

d) A wrong initial diagnosis.

e) Dietary non-compliance.

f) Progression to chronic pancreatitis

Objective: Recognize the causes of persistent diarrhea in celiac disease.

An 8-month history of unintentional weight loss with bulky pale stools would point to malabsorption. The stool findings suggest steatorrhea, but with no blood or pus. There is dimorphic blood picture; note the hypochromic RBCs with increased MCV. The positive serum testing for IgG anti-tissue transglutaminase would point out towards celiac disease; the histopathological examination of the duodenal biopsy confirms this observation. The negative serum IgA anti-endomysial antibodies reflect an associated selective IgA deficiency.

The physician most likely advised a gluten-free diet and this was the cause of the initial improvement; this eliminates the possibility of wrong initial diagnosis. Non-compliance with this form of diet is the commonest cause of treatment failure/relapse of celiac patients. The other stems are distractions. ***(Correct Answer: d.)***.

6) An otherwise healthy 19-year-old man visits the doctor's office. He states that he gets jaundiced every now and then. You do some blood tests for him. The results are as follows:

Serum total bilirubin:	3.1 mg/dl
Serum indirect bilirubin:	2.9 mg/dl
ALT:	11 iu/L
Serum alkaline phosphatase:	73 iu/L
Serum albumin:	4.1 g/L
Hb:	13.5 g/dl
Blood urea:	38 mg/dl
Urine examination:	unremarkable

Blood film: normochromic normocytic red blood cells, no immature cells, and adequate platelets

What is the diagnosis?

a) Hereditary spherocytosis

b) Choledochal cyst

c) Chronic hepatitis C infection

d) Gilbert's syndrome

e) Crigler-Najjar syndrome, type I

Objective: Review the causes of isolated indirect hyperbilirubinemia.

There is recurrent jaundice of indirect (pre-hepatic) hyperbilirubinemia. The rest of the liver function testing is unremarkable, as are the hemoglobin and blood film. This would fit Gilbert's syndrome. If the hemoglobin is low (or the blood film is abnormal), think of chronic/fluctuating hemolysis. Choledochal cyst would produce cholestasis with direct hyperbilirubinemia and elevated serum alkaline phosphatase. Chronic hepatitis C infection would result in hepatocellular damage (with raised serum liver aminotransaminases) and mixed hyperbilirubinemia; a clue towards this infection is usually provided in the question (such as intravenous drug abuse, hemophilia, hemodialysis…etc.). Isolated indirect hyperbilirubinemia also occurs in Crigler-Najjar syndrome, but type I disease is associated with severe jaundice and neurologic impairment due to bilirubin encephalopathy that can result in permanent neurologic sequelae (kernicterus).

Most cases of Gilbert's syndrome are usually diagnosed in young adults who present with mild indirect hyperbilirubinemia. The syndrome is rarely diagnosed before puberty, when alterations in sex steroid concentrations affect bilirubin metabolism and this would lead to increased plasma bilirubin concentrations. It is mainly observed in males, possibly due to a relatively higher level of daily bilirubin production. ***(Correct Answer: d.)***.

7) A 54-year-old man visits the physician's office because of watery diarrhea. His bowel opens 5-8 times a day. This has been occurring over the past few months. Otherwise, he is healthy. His GP did sigmoidoscopy for him, 1 week ago, and the result was normal. He denies blood or pus in his stool. After ordering some investigations, you get these findings:

Hemoglobin:	11.1 g/dl
WBCs:	7.4 x 109/L
Platelets:	210 x 109/L
Blood urea:	42 mg/dl
Serum creatinine:	1.0 mg/dl
Serum potassium:	2.9 mEq/L
Serum sodium:	130 mEq/L
Fasting glucose:	5.1 mmol/L
Urine:	normal
3-day stool fat:	10 mmol/day
Stool:	no pus cells or RBCs, no parasites
Stool osmolar gap:	40 mosm/L

What is the cause of this man's diarrhea?

a) Crohn's disease

b) Bile salt diarrhea

c) Chronic giardiasis

d) Celiac disease

e) Pancreatic VIPoma

Objective: Recall the diagnostic approach of secretary diarrhea.

Chronic watery diarrhea with hypokalemia, normal stool osmolar gap (<50 mosm/L), no blood, no pus, and parasites is highly suggestive of secretary type of diarrhea. The differential diagnosis is between pancreatic VIPoma (so-called pancreatic cholera) and the rectal villous adenoma; the latter was excluded by the normal sigmoidoscopy. Pancreatic somatostinoma usually results in steatorrhea (note the normal 3-day fat collection in this patient and the absence of diabetes). Acute secretary diarrheas are usually infective (e.g., travelers' diarrhea).

High stool osmolar gap (>50 mosm/L) indicates the presence of an osmotically active substance in stool, such as sorbitol. Osmotic diarrheas characteristically improve with fasting, while the secretary diarrheas are not influenced by fasting. ***(Correct Answer: e)***.

8) A 32-year-old man from South East Asia comes for a routine health insurance assessment. He denies symptoms and his past medical history is unremarkable. His BMI is 18 Kg/m^2. These are the results of his investigations:

Hemoglobin:	14 g/dl
WBCs:	$6.2 \times 10^9/L$
Blood urea:	29 mg/dl
Serum creatinine:	0.9 mg/dl
Serum total bilirubin:	0.8 mg/dl
ALT:	10 iu/L
Serum alkaline phosphatase:	76 iu/L
Serum LDH:	270 iu/L
Anti-HBc IgG:	positive
Anti-HBc IgM:	negative
HBs antigen:	negative
Anti-HBs IgG:	positive
Urine:	normal
Chest X-ray:	bi-apical fibrotic changes

What would you do next?

a) Start interferon alpha

b) Advise lamivudine therapy

c) Counseling for HIV testing

d) Do serum alpha fetoprotein

e) Re-assurance

Objective: Review serology of hepatitis B infection.

In addition to being a little bit thin, this man shows an evidence of old healed pulmonary tuberculosis, as well as serological recovery from acute hepatitis B infection. The absence of HBs antigen, normal liver functions, and positive anti-HBs and IgG anti-HBc would favor the latter notion (i.e. recovery from an acute hepatitis B viral infection). There is no need to start anti-HBV therapy; he does not demonstrate any high-risk behavior for HIV infection. In patients with cirrhosis, serial serum testing for alpha-fetoprotein and ultrasound of the liver may be used in screening for the development of hepatocellular carcinoma; our patient does not belong to this category. Apart from reassurance, actually, he needs testing for "latent" tuberculosis! ***(Correct Answer: e)***.

9) A 33-year-old man visits the Emergency Room complaining of right hypochondrial pain and severe nausea. He is HIV-positive, and his up-to-date CD4+ cell count is $92/mm^3$. You found no fever. There is tenderness in the right hypochondrium and epigastrium, as well as hepatomegaly. You do the following tests:

Serum total bilirubin:	3.0 mg/dl
AST:	60 iu/L
ALT:	64 iu/L
Alkaline phosphatase:	770 iu/L
Serum albumin:	3.2 g/dl
Abdominal ultrasound:	No gallstones; common bile duct diameter is 13 mm, no stones.

What would you do next?

a) Endoscopic retrograde cholangiopancreatography (ERCP)

b) Abdominal CT scan with contrast

c) Percutaneous transhepatic cholangiography (PTC)

d) Esophagogastroduodenoscopy

e) Liver biopsy

Objective: Review the diagnostic approach of AIDS cholangiopathy.

AIDS cholangiopathy is a syndrome of biliary obstruction resulting from infection-related strictures of the biliary tract. The organism most closely associated with AIDS cholangiopathy is *Cryptosporidium parvum*; other pathogens that have been identified include *Microsporidium*, *cytomegalovirus* (CMV), and *Cyclospora cayetanensis*. Involvement of large intra-hepatic ducts is usually associated with *C. parvum* and CMV infections.

AIDS cholangiopathy is usually seen in patients with a CD4+ count well below 100/mm^3 and may be their presenting manifestation. Affected patients typically present with right upper quadrant and epigastric pain and diarrhea; fever and jaundice are less common, occurring in 10 to 20% of patients. The severity of the abdominal pain varies with the biliary tract lesion. Severe abdominal pain is indicative of papillary stenosis, while milder abdominal pain is usually associated with intra-hepatic and extra-hepatic sclerosing cholangitis without papillary stenosis. The diagnosis of AIDS cholangiopathy is usually made by ERCP; combined papillary stenosis and sclerosing cholangitis is the commonest finding (60% of patients). The survival of patients is not affected by cholangiopathy, since the mortality rate is primarily determined by the natural history of AIDS. ***(Correct Answer: a)***.

10) A 68-year-old man visits the Emergency Room because of dizziness and severe painless lower GIT bleeding for 2 hours. You find postural hypotension. When you do per rectal examination, you detect blood in the rectal vault, but no mass is found. You put a nasogastric tube and aspirate bile only.

What is the reason of this man's presentation?

a) Bleeding diverticular disease of the colon

b) Acute ischemic colitis

c) Proctitis phase of ulcerative colitis

d) Angiodysplasia of the colon

e) Sigmoid volvulus

Objective: Recall the commonest causes of lower GIT bleeding in elderly people.

Colonic carcinoma is the most common source of lower gastrointestinal blood loss, while diverticular bleeding is the most common cause of brisk hematochezia (maroon or bright red blood), accounting for 30 to 50% of cases of massive rectal bleeding. Bleeding will occur in 15% of patients with diverticulosis, being massive in approximately one-third. Unlike diverticulitis, which occurs primarily in the left colon, the right colon is the source of diverticular bleeding in 50 to 90% of patients. Patients have few, if any, abdominal symptoms, which is a reflection of the non-inflammatory pathogenesis of the bleeding. Angiodysplasia accounts for another 20 to 30% of cases of hematochezia and may be the most frequent cause in patients over the age of 65. It is also one of the major causes of gastrointestinal bleeding, particularly recurrent bleeding, in patients with end-stage renal disease. ***(Correct Answer: a)***.

11) A 43-year-old woman has noticed painful swallowing at the level of mid-sternum over the past 10 days. She is HIV-positive. She does not take her highly active anti-retroviral therapy. She has bloody diarrhea and impaired vision. Her CD4+ count is 55/mm^3. Mouth examination is unremarkable. You do esophagoscopy and find a single shallow ulcer of 13 cm in maximum diameter, with a normal-looking surrounding mucosa.

What is the cause of this woman's odynophagia?

a) Herpes simplex

b) Candida

c) HIV itself

d) CMV

e) Surreptitious drug abuse

Objective: Review characteristic features of infectious esophagitis in HIV patients.

CMV esophagitis usually results in multiple discrete ulceration; sometimes these ulcers coalesce to form a giant ulcer. CMV esophagitis is usually part of a more disseminated CMV disease; this woman also has bloody diarrhea (CMV colitis) and visual impairment (which might result from CMV retinitis). The most common gastrointestinal sites of CMV involvement are the esophagus and the colon. The oral mucosa is characteristically normal in CMV esophagitis, while oral lesions usually co-exist with *Herpes simplex* and *Candida* esophagitis. Some HIV-infected patients develop giant esophageal ulcers but no pathogen is isolated; these ulcers must be differentiated from giant CMV ulcers and large ulcers associated with HIV per se. ***(Correct Answer: d).***

12) A 34-year-old man is brought to the Emergency Room by his wife. He has been experiencing severe upper abdominal pain and nausea over the past day. He reports a similar but milder pain, 2 months ago. Apart from insulin, he denies taking other medications or ingesting alcohol. He has poorly controlled type I diabetes. You find hypotension, tachycardia, epigastric tenderness, and diminished bowel sounds. There is a papular rash on both knees. WBC is 13,000/mm^3, blood urea 40 mg/dl, random blood glucose 410 mg/dl, and ALT 12 iu/L. Serum amylase is within its normal reference range. The abdominal ultrasound's report mentions pancreatic swelling.

What is the cause of this abdominal pain?

a) Chronic alcoholic pancreatitis

b) Acute gallstones pancreatitis

c) Acute viral pancreatitis

d) Acute pancreatitis secondary to hypertriglyceridemia

e) Prolonged hypercalcemia

Objective: Review the causes of acute pancreatitis and their clues.

This patient has acute pancreatitis. Alcohol ingestion and gallstones are responsible for about 70% of acute pancreatitis cases. However, the poorly controlled diabetes and the papular knee rash in this man are well compatible with prolonged/severe secondary hypertriglyceridemia. Serum triglyceride concentrations >1000 mg/dl can precipitate attacks of acute pancreatitis, although the pathogenesis of inflammation in this setting is unclear. Hypertriglyceridemia accounts for 1.3 to 3.8% of cases of acute pancreatitis. The clinical manifestations of hypertriglyceridemia-associated pancreatitis are similar to those seen with other causes of acute pancreatitis with abdominal pain, nausea, and vomiting being the major complaints. The serum amylase may not be elevated in such cases, at presentation, for reasons that are not well understood, however. ***(Correct Answer: d).***

13) A 48-year-old woman is referred to the gastroenterology outpatient clinic. She has been experiencing intermittent abdominal pain since 3 months, that is poorly responsive to spasmolytic agents. Abdominal ultrasound, which was done 1 month ago, was unremarkable. She was diagnosed with irritable bowel syndrome. However, you order abdominal CT scan and this detects a cyst, 5 x 4 cm in maximum diameter, in the pancreas.

What would you do next?

a) Endoscopic drainage of the cyst

b) Wait and see; repeat the CT scan after 3 months

c) Measure serum amylase

d) Surgical removal of the cyst

e) Give pancreatic enzymes

Objective: Review the management of pancreatic cysts.

Cystic lesions of the pancreas may be divided pathologically into retention cysts, pseudocysts, and cystic neoplasms. Simple (true or retention) cysts of the pancreas are small, developmental, fluid-containing spaces lined by normal duct and centroacinar cells. They are usually incidental findings, which are of no clinical significance and can be left untreated. Pseudocysts of the pancreas develop as a result of pancreatic inflammation and necrosis. They may be single or multiple, small or large, and can be located either within or outside of the pancreas. Four types of cystic neoplasms of the pancreas have been described: mucinous cystadenoma/ cystadenocarcinoma; mucinous duct ectasia (intraductal papillary mucinous neoplasm); serous cystadenoma; and papillary cystic neoplasm. Mucinous cystadenoma/cystadenocarcinoma is the most common cystic neoplasm. It typically occurs in middle-aged women, and is usually located in the body or tail of the pancreas. Most of these mucinous tumors are malignant at the time of diagnosis; even those which appear to be benign have a high potential for malignant change. The presence of painful pancreatic cyst in a patient without a history of acute pancreatitis and who is not at risk of acute pancreatitis should be prompt the physician consider neoplastic masses, not a pseudocyst. This patient's cyst should be excised surgically and its histology should be defined. ***(Correct Answer: d).***

14) A 51-year-old woman visits the physician's office because of a 2-day history of painful hematuria. The physician orders abdominal ultrasound and found 2 small stones in the left renal pelvis. However, the ultrasonography report also mentions 4 gallstones, the largest is 2 x 1.5 cm in maximum diameter, without dilatation of the common bile duct. The patient denies any history of right hypochondrial pain or dyspepsia. You prescribe treatment for her renal stones.

What would you suggest for the gallstones?

a) Elective laparoscopic cholecystectomy

b) Emergency cholecystectomy

c) Ursodeoxycholic acid

d) Pre-operative ERCP

e) Advise against doing cholecystectomy and do regular follow-up

Objective: Review the management of asymptomatic gallstones.

Gallstone disease is one of the most common and costly of all digestive diseases. The routine use of ultrasonography for the evaluation of abdominal pain, pelvic disease, and abnormal liver function tests…etc. has led to the identification of incidental gallstones in many patients. The majority of these patients have no symptoms attributable to the gallstones; however, approximately 20% will become symptomatic at up to 15 years of follow-up. Most guidelines recommend against prophylactic cholecystectomy in most patients with asymptomatic gallstones. Possible exceptions include patients who are at increased risk for gallbladder carcinoma or gallstone complications, in whom prophylactic cholecystectomy or incidental cholecystectomy at the time of another abdominal operation can be considered. ***(Correct Answer: e).***

15) A 46-year-old man is brought to the Emergency Department after developing severe hematemesis. He has Child's B liver cirrhosis. You resuscitate him and do esophagogastroscopy. You find bleeding esophageal varices but no gastric ulcers or erosions. You intend to intervene endoscopically to stop this hemorrhage.

What would you choose?

a) Esophageal trans-section

b) Sclerotherapy

c) Local injection of adrenalin

d) Band ligation

e) Don't intervene for the time being

Objective: Review the management of esophageal variceal bleeding.

The patient has been resuscitated and the cause of this hemorrhage has been identified; the hemorrhage must be stopped. Sclerotherapy utilizes a sclerosant agent, such as sodium tetradecyl sulfate or ethanolamine oleate. The injection of these into a varix would induce thrombosis. However, this form of therapy has lost its popularity because of its adverse effects, e.g., retrosternal tightness and pain, esophageal ulceration (which may be large and usually heals within 3 weeks spontaneously), perforation (which may be delayed for 1-4 weeks), stricture formation, and future dysphagia. These complications may afflict up to 35% of patients. In band ligation, the operator sucks the varix into the ligator cap and deploys a rubber band around its neck. This would cause plication of the varix and its submucosa ending with fibrosis and obliteration of the varix. The complications of this form of therapy include retrosternal pain, superficial ulceration, and perforation; their rate is 2-19%. Many studies showed that band ligation results in better control of variceal hemorrhage with fewer rate of complications when compared with sclerotherapy. Band ligation is the intervention of choice in this patient. Stem "a" is an open surgery which is done under general anesthesia, and stem "c" is used in gastric ulcer bleeding in some patients. ***(Correct Answer: d.).***

Chapter 10

Nephrology
10 Questions

This page was left intentionally blank

1) A 6-year-old boy is brought by his mother to your office. She says that her son was diagnosed with nephrotic syndrome, 3 months ago, after doing some investigations. She insists that her son's face is still puffy and his legs are swollen with little improvement since his treatment has been initiated with prednisolone 60 mg per day. She gives the medication to him every day and has never missed a dose. She declines insect bites or allergies. The child's uncle died of polycystic kidney disease and his grandmother has diabetic nephropathy. The boy's mother is anxious and desperate for your help. Examination reveals gross Cushingoid habitus, periorbital puffiness, and prominent pitting leg and sacral edema. Blood pressure is 150/100 mmHg; it was 110/60 mmHg at the time of diagnosis. Urine shows ++++ proteinuria, no RBCs, and no casts. Serum cholesterol is raised and there is hypogammaglobulinemia.

What would you do next?

a) Increase the dose of prednisolone

b) Continue the same dose for another 3 months

c) Add cyclophosphamide

d) Do renal biopsy

e) Refer for kidney transplantation

Objective: Review the management of steroid-resistant nephrotic syndrome.

This boy has obviously failed his medical treatment and iatrogenic Cushing's syndrome has ensued. Urine is grossly proteinuric, and he is clinically edematous. The boy was diagnosed with nephrotic syndrome by the combination of his clinical features and lab tests, and was given a treatment for a presumed minimal changed disease. Pediatric nephrologists avoid doing renal biopsy because minimal change disease is the commonest cause of nephrotic syndrome in children, and they do renal biopsy whenever the clinical picture is atypical (hypertension at the time of diagnosis, hematuria, impaired renal function…etc.), or the syndrome is unresponsive to steroids (as in our boy). this kid's hypertension is due to his high-dose prednisolone. Generally, 90% of children with suspected minimal change disease will respond to steroids within 4 weeks. The rest will usually respond after a further 2-4 weeks' period.

Those who don't respond after this period may try methyl prednisolone pulses and should be reviewed after 1 week If urine is still grossly proteinuric, the patient is labeled as having steroid-resistant nephrotic syndrome, and renal biopsy should be done. Adult nephrologists prefer to do renal biopsy at the time of presentation because they believe that knowing the etiology of nephrotic-range proteinuria would help decide the best treatment plan by finding an unexpected diagnosis through histological examination. ***(Correct Answer: d)***.

2) A 63-year-old accountant develops lassitude, anorexia, and drowsiness. He has chronic stable angina, for which he underwent coronary angiography 4 weeks ago. The angiography showed critical stenosis of the proximal right coronary artery, that was successfully ballooned and stented with a drug-eluting stent. His daily medications are metoprolol, amlodipine, simvastatin, and aspirin. Examination shows blood pressure 160/110 mmHg, pulse rate 90 beats/minute, clear chest, and unremarkable precordium. There are bluish reticulated rashes on both ankles and blue right big toe. Today, his blood tests show blood urea 140 mg/dl, serum creatinine 3.4 mg/dl, and low serum complements. One month ago, the above results were as follows: blood urea 33 mg/dl and serum creatinine 1.1 mg/dl.

What does the man have developed?

a) Renal artery dissection

b) Renal vein thrombosis

c) Contrast nephropathy

d) Cholesterol atheroembolic disease

e) Polyarteritis nodosa

Objective: Recall atheroembolic disease.

Obviously, this man has renal failure, which has developed over a matter of 4 weeks. Contrast nephropathy occurs *within* 24 hours of undergoing invasive vascular procedures using a contrast material (coronary angiography, aortography…etc.). The clinical presentation fits polyarteritis nodosa but the preceding history of coronary angiography makes cholesterol atheroembolic disease the most likely diagnosis (which develops over days, weeks, and even months). In addition, hypocomplementemia is not seen in polyarteritis nodosa but is consistent with cholesterol atheroembolic disease. Bilateral renal artery dissection (upon doing angiography) is rare and presents with *acute* renal shutdown. Renal artery thrombosis presents with gross hematuria and rapid renal impairment. There is no specific treatment for cholesterol atheroembolic disease. ***(Correct Answer: d.).***

3) A 64-year-old retired editor presents with bilateral flank heaviness. He is hypertensive for the past 4 years but his blood pressure is well-controlled with daily 80 mg nadolol. He denies fatigue, anorexia, weight loss, or difficult urination. His mother died after many years of hemodialysis because of a certain kidney disease. He takes oral vitamins every now and then. On physical examination, the man looks healthy, is normotensive, and has unremarkable chest and heart. There are bilateral asymmetric non-tender flank masses. Abdominal ultrasound shows many cysts in each kidney, both of which are enlarged, but neither stones nor hydroureter are found. Tests are as follows: Hb 12 g/dl, ESR 16 mm/hour, blood urea 60 mg/dl, serum creatinine 1.9 mg/dl, serum calcium 8.9 mg/dl, serum potassium 4.2 mEq/L, and few amorphous oxalate crystals are found on urine examination.

How would manage this man?

a) Arrange for hemodialysis

b) Do abdominal CT scan

c) Advise for allogenic kidney transplantation

d) Refer for peritoneal dialysis

e) Continue his current treatment

Objective: Review the management of adult polycystic kidney disease.

The overall clinical, lab, and imaging features are consistent with adult polycystic kidney disease (APKD), with mild renal impairment, normal serum electrolytes, and normal hemoglobin. Up 50% of APKD patients never need dialysis; those who need dialysis are usually in their 6th decade. Proper control of blood pressure, aggressive treatment and prevention of UTI, and avoidance of nephrotoxic medications are all that is required for the time being. The patient has no end-stage renal disease to arrange for renal replacement therapy. The ultrasound study suffices for the diagnosis and abdominal CT scanning would add nothing useful. ***(Correct Answer: e).***

4) A 12-year-old boy was referred from the pediatrics department as a newly diagnosed case of diabetes mellitus. You do these tests:

Hemoglobin:	13 g/dl
Blood urea:	34 mg/dl
Serum creatinine:	0.7 mg/dl
Serum potassium:	3.0 mEq/L
Serum sodium:	132 mEq/L
Serum bicarbonate:	14 mEq/L
Serum chloride:	110 mEq/L
Urine:	++ sugar, no protein, no RBCs
Plasma glucose:	5.0 mmol/l
Serum uric acid:	1.7 mg/dl

What is the diagnosis of this boy's illness?

a) Maturity onset diabetes of the young (MODY) type III

b) Lag storage disease

c) Fanconi syndrome

d) Diabetic ketoacidosis

e) Minimal change disease

Objective: Recall renal tubular defects.

No clinical features were provided. When you analyze the patient's data, you will find normal anion gap metabolic acidosis and normal renal function. The positive urinary sugar testing (with normal blood sugar) and the very low serum uric acid would label the patient as having a generalized proximal renal tubular defect (type II renal tubular acidosis). He has no diabetes (including MODY). Nephrotic-range proteinuria is found in stem "e", and stem "d" should have high blood glucose and positive urinary ketones. Lag storage disease, or alimentary hyperglycemia, is s very different situation. ***(Correct Answer: c.).***

5) A 30-year-old secretary comes for a check-up. She was lastly seen by her GP 4 years ago. She was diagnosed with type I diabetes mellitus at the age of 14 years. She has been using insulin injections since then. She says that her fingers and feet numb and she feels tired most of the time in spite of having good sleep at night. Her duties at work are demanding and she misses a meal every now and then. She documented frequent "hypos." She denies chest pain or breathlessness. She neither smokes nor drinks alcohol. She lives with her husband in a 2-story house but they have no children. Her daily medications are soluble and isophane insulin and tonics. Examination reveals a thin woman with BMI of 18 Kg/m^2, pulse rate of 86 beats per minute, and blood pressure of 160/105 mmHg, and pitting ankle edema. There is diminished pinprick and touch sensations in a stocking pattern with planter callosities. The fundi have background retinopathy. The neck, chest, and abdomen are unremarkable. Serum triglyceride is 310 mg/dl, serum cholesterol is 240 mg/dl, serum LDL-cholesterol is 140 mg/dl, and urine is positive for protein and sugar. Blood urea is 85 mg/dl and serum creatinine is 1.9 mg/dl. HBA1c is 9%.

All of the following are part of your management plan, *except*:

a) Optimize her glycemic control

b) Add a statin

c) Do renal biopsy

d) Prescribe enalapril

e) Give gabapentin at night

Objective: Review the management of diabetic nephropathy.

Diabetes mellitus is disease that can result in many complications, which are responsible for most of its morbidity and mortality. This woman has long-standing poorly controlled type I diabetes, diabetic retinopathy, peripheral neuropathy, and nephropathy; all of them need to be addressed and managed appropriately. In general, this woman needs:

1. Optimization of her blood sugar control. Her insulin regimen needs to be re-assed and changed accordingly.

2. Proper control of blood pressure. Hypertension is uncommon in young patients with type I diabetes (while in type II disease hypertension is common), and its presence should prompt the physician search for diabetic nephropathy. ACE inhibitors are excellent options to treat both, the hypertension and the kidney disease. The target blood pressure in this patient is <130/80 mmHg. Diabetes mellitus is a coronary artery disease equivalent. Serum LDL must be maintained below 100 mg/dl. This woman's lipid profile is highly atherogenic and needs to be reduced using a statin and dietary plans to bring down serum cholesterol and LDL-cholesterol back to their target. Raised serum triglyceride is usually seen in poorly controlled diabetes and usually becomes lower with proper glycemic control. Total cholesterol should be maintained below 200 mg/dl while serum triglyceride should be less than 150 mg/dl.

3. Symptomatic treatment of her peripheral neuropathy with protection of the feet from trauma and infections. Gabapentin can be used to alleviate her neuropathic pain.

4. The fundi should be examined with pupillary dilatation annually, and referral to the ophthalmologist should be done at the proper stage.

5. Diabetic nephropathy is managed by proper control of blood sugar and blood pressure and the use of ACE inhibitors.

The cause of her renal disease can be clearly ascribed to diabetes; no need for doing renal biopsy. If the clinical history or lab findings are suggesting a non-diabetic kidney disease (e.g., hematuria, palpable purpura…etc.), then kidney biopsy should be taken. ***(Correct Answer: c)***.

6) A 59-year-old woman visits the Acute and Emergency Department because of fever. She also has chills, painful micturition, and frequency. Her symptoms started abruptly yesterday. She has adult polycystic kidney disease and hypertension. She takes captopril, 50 mg, three times daily. The patient admits to losing 11 Kg unintentionally over the past 4 months. She is thin, her weight is 55 Kg, and she looks ill. You find a temperature of 39.5 ^{o}C, blood pressure 90/50 mmHg, pulse rate 108 beats per minute, and respiratory rate 22/minute. Both costophrenic angles are tender. Serum creatinine is 1.2 mg/dl, and general urine examination shows full-field pus cells and bacteruria. You give intravenous gentamycin and ampicillin with antipyretics. You intern suggests dose modification/adjustment for these antibiotics.

Why does he suggest that?

a) Because of hypertension

b) Because of systemic sepsis

c) Because of microbial resistance

d) Because of renal cysts' infection

e) Because of reduced glomerular filtration rate

Objective: Recognize that the GFR might well be reduced in spite of having a normal serum creatinine.

If we correct the serum creatinine with the patient's age and body weight, we will find that the glomerular filtration rate (GFR) is greatly reduced. Aminoglycosides need major dose reduction in this patient. Approximately, half of women with serum creatinine level of greater than 1.0 have a reduced GFR. The mean rate of decline in creatinine clearance is 0.75 mL/min/year, and is greater in patients with hypertension. The overall clinical picture is pointing out towards bilateral pyelonephritis rather than cyst infection; therefore, this patient's regimen has included ampicillin. Penicillins penetrate renal cysts *poorly* while quinolones, trimethoprim/sulfamethoxazole, and chloramphenicole have a good penetration level. ***(Correct Answer: e).***

7) A 19-year-old intravenous drug addict man visits the physician's office. He has lethargy and nausea. You run some tests and find out that the blood urea is 50 mg/dl and serum creatinine is 2.1 mg/dl. A spot urinary albumin/creatinine ratio is 12. The right kidney's length is 12 cm while the left one is 11.5 cm. You test him for hepatitis B and C and the results turn out to be negative. He agrees to undergo HIV testing; this is positive. You consider renal biopsy.

What is the likely histopathological diagnosis of this man?

a) Membranous nephropathy

b) Minimal change disease

c) Normal findings

d) Diffuse mesangial expansion

e) Collapsing focal segmental glomerulosclerosis

Objective: Recall the pathophysiology of HIV-nephropathy.

Renal disease is a relatively common complication in patients infected with HIV. A collapsing form of focal glomerulosclerosis (FGS) has been considered the primary form of HIV nephropathy; up to 60% of renal biopsies showing this lesion. HIV-associated FGS occurs in 2 to 10% of HIV-infected patients, usually associated with a high viral load. This patient has developed nephrotic-range proteinuria on the background of impaired renal function and *normal* kidney size. ***(Correct Answer: e).***

8) A 17-year-old athlete is referred to your office for further evaluation of dipstick-positive urinary testing for blood, which was done as part of sports physical. The patient denies symptoms and his past medical history is unremarkable. Careful inquiry does not reveal any family history of kidney diseases. He has a well-built physique, blood pressure 142/99 mmHg, and a body weight of 81 Kg. You repeat his urine examination and find pH 5, ++ protein and blood, no pus cells. Blood tests are pending. Abdominal ultrasound examination is unremarkable.

What is the next best step in the management of this young man?

a) Renal biopsy

b) Abdominal CT scan with contrast

c) MRA of both renal arteries

d) Urine culture for acid fast bacilli

e) Wait and see; repeat his urine examination after 2 weeks

Objective: Review the diagnostic approach of unexplained hematuria/proteinuria.

The occurrence of significant proteinuria and hematuria in a healthy-looking young athlete with borderline high blood pressure and normal renal ultrasound is abnormal and should prompt the physician assess the upper urinary tract, and that is the kidneys, by doing renal biopsy. If no diagnosis is apparent from the history, urinalysis, radiologic imaging tests, or cystoscopy, then the most likely causes of persistent isolated hematuria are mild glomerulopathy (such as thin basement membrane nephropathy) and predisposition to stone disease, particularly in young and middle-aged patients. ***(Correct Answer: a)***.

9) A 21-year-old alcoholic man was found on the floor of his backyard and was brought to the Emergency Department after then. He is comatose and dehydrated. You run a battery of investigations and the results are: Hb 13.2 g/dl, blood urea 51 mg/dl, serum creatinine 1.6 mg/dl, serum creatine phosphokinase 14,500 iu/L, random plasma glucose 88 mg/dl. Urine is reddish and is dipstick positive for blood (++++) and protein (+). Microscopic examination of this urine sample shows 1-2 RBCs and 1-3 pus cells per high power field.

What does urine examination suggest?

a) Alcoholic ketoacidosis

b) Methanol poisoning

c) Myoglobinuria

d) Post-obstructive azotemia

e) Acute interstitial nephritis

Objective: Recall predisposing factors for rhabdomyolysis and its diagnostic approach.

The presence of strongly positive urine dipstick testing for blood with absence of red blood cells on urine microscopy is suggestive of myoglobinuria or hemoglobinuria. This, together with the very high serum level of creatine phosphokinase, would indicate rhabdomyolysis and myoglobinuria, which have complicated acute alcoholic intoxication and prolonged immobilization. Myoglobin is a heme-containing respiratory protein and is released from damaged muscle in parallel with creatine phosphokinase. Myoglobin is a monomer that is not protein-bound and is therefore rapidly excreted in the urine, often resulting in the production of red to brown urine. Because of rapid excretion, myoglobin does not produce a change in plasma color (unlike hemoglobin in intravascular hemolysis) unless renal failure limits myoglobin excretion. Myoglobinuria may be absent in patients with established renal failure or those who present late in the course. In addition, myoglobin is cleared from the plasma more rapidly than creatine phosphokinase. Thus, it is not unusual for creatine phosphokinase levels to remain elevated in the absence of myoglobinuria. ***(Correct Answer: c)***.

10) A 35-year-old man is referred from the surgical department for further management. He underwent renal transplantation, 1 week ago. The patient is still oliguric and the surgeon plans to do dialysis. You tell the surgeon that there are many potential causes for this delayed graft function (DGF).

What is the commonest cause of this DGF?

a) Urinary tract obstruction by hematoma

b) Hyper-acute graft rejection

c) Post-ischemic acute tubular necrosis

d) Thrombosis of the renal vein

e) Atheroemboli

Objective: Recall the causes of delayed graft function.

Renal failure persisting after transplantation is called delayed graft function (DGF). DGF generally refers to oliguria *or* the requirement for dialysis in the first week post-transplantation. Less than 5% of kidneys with DGF never function (primary non-function). The causes of DGF (which has a major impact on graft survival) are post-ischemic acute tubular necrosis (ATN), hyper-acute and acute antibody mediated rejection, accelerated rejection superimposed on ischemic ATN, urinary tract obstruction due to ureteral necrosis with urinary leak or to hematoma, atheroemboli, and thrombosis of the renal artery or vein. Post-ischemic ATN is the commonest cause. ***(Correct Answer: c).***

Chapter 11

Dermatology
5 Questions

This page was left intentionally blank

1) A 19-year-old female is referred by her GP because of an itchy skin condition. The patient has hyperpigmented lichenified papular rash over the face, as well as the antecubital and popliteal fossae. There are excoriation marks but there are little oozing and scaling. She has had this rash since the age of 1 year but their locations have been changing. Her older brother has intermittent asthma. She is currently sexually inactive, and her menstrual cycle is irregular. HIV status is negative.

Which one of the following is *true* with respect to this woman's illness?

a) Progression to wide-spread flaccid skin bullae is common

b) Topical tacrolimus can be used for the facial rash

c) Super-infections of the rash can induce remission

d) Family history of the same rash is unusual

e) Systemic glucocorticoids are the first-line agents in the treatment

Objective: Review the management of atopic dermatitis.

This woman displays features of atopic dermatitis. Personal or family history of atopy is common. The skin loses its barrier function and becomes vulnerable to a variety of bacterial, viral, and fungal infections, which will exacerbate the skin lesions; therefore, skin infection is both a trigger and a complication of atopic eczema. Atopic eczema has no correlation with pemphigus. The cornerstone in the treatment (in addition to avoiding skin irritants and proper treatment of superadded infections) is the combination of topical emollients and topical steroids; topical calcineurin inhibitors can added (tacrolimus, pimecrolimus) when the rash involves the face or skin folds. Systemic steroids are used when topical measures fail. ***(Correct Answer: b).***

2) A 65-year-old journalist from Australia comes to see you because of having had many skin lesions on his scalp for the past few years. He is bald and the lesions are multiple small red areas with scales. The lesions neither itch nor bleed. There are no areas of ulcerations. He is anxious and says that he knows someone with a skin lesion and his GP told him that it is minor infection but it turned out to be a skin malignancy. He is desperate for your help.

What are you going to tell this man?

a) The lesions have high risk of progression to malignant melanoma

b) Skin biopsy will show full thickness dysplastic changes

c) Lesions may involute spontaneously

d) Ulceration means healing

e) Liquid nitrogen is contraindicated as a treatment modality

Objective: Recall the natural history of actinic keratosis.

The characteristics of the rash is consistent with actinic keratosis; focal areas of partial dysplasia in the dermis due to the effect of sun light. They are common in white people living for a long time in the tropics. The lesions are mainly seen in sun-exposed areas. The progression to skin squamous cell cancer is extremely low. Most actinic keratoses do not progress to skin squamous cell cancer; the risk of malignant transformation of an actinic keratosis within one year is approximately 1 in 1000. On the other hand, approximately 60% of skin squamous cell cancers probably arise from actinic keratoses. Some of the lesions may involute without any intervention. Ulceration, bleeding, or pain should prompt the physician think of squamous cell cancer development. Small discrete lesions can respond to local application of liquid nitrogen. Multiple confluent lesions can be treated with topical 5-FU. Those, which don't respond to topical measures, may be removed by curettage or even excision; resistant lesions should be re-evaluated. ***(Correct Answer: c).***

3) A GP refers to you this 26-year-old man who has had many pustules on his palms and soles for the past few months. You examine the patient and diagnose localized palmo-planter pustular psoriasis.

What is the first line treatment for this subtype of psoriasis?

a) Topical emollients

b) Topical tar

c) Oral prednisolone

d) Oral cyclosporine

e) Oral acitretin

Objective: Review the treatment of psoriasis subtypes.

Localized pustular psoriasis of the palms and soles is difficult to treat. Approaches include potent topical glucocorticoids and topical bath PUVA. Systemic retinoids, such as acitretin, are particularly effective in the treatment of pustular psoriasis and are the first line therapy. ***(Correct Answer: e).***

4) You are following up this 45-year-old man from Australia who was treated for stage I malignant melanoma of the right shin. The lesion underwent surgical resection 6 months ago. You notice a re-growth in the same area. You do many investigations and diagnose local recurrence but you fail to detect any evidence of the disease elsewhere.

Which one of the following is the best treatment option for this man?

a) Surgical resection of the locally recurred tumor

b) Neoadjuvant chemotherapy with dacarbazine

c) Interleukin-II

d) Interferon gamma

e) Amputation of the right leg

Objective: Review the management of local recurrence of skin's malignant melanoma.

Local recurrences of melanoma following initial therapy may occur by a variety of mechanisms and generally are a harbinger of disseminated disease. Although the long-term prognosis for most patients is poor, aggressive therapy is warranted to minimize morbidity and potentially to increase survival. For patients with a local recurrence and *no* other evidence of disease, surgical resection is the best option; lymphatic mapping and sentinel lymph node biopsy should be done next, with completion lymph node dissection if positive lymph nodes are identified. ***(Correct Answer: a).***

5) A 33-year-old man presents with severely pruritic polygonal violet flat-topped papules on his penis. You tell him that his disease is called lichen planus.

How would you treat?

a) Oral tacrolimus

b) PUVA

c) Topical triamcinolone

d) Cryotherapy

e) Topical 5-fluorouracil

Objective: Review the treatment of lichen planus and its varieties.

Genital lesions of lichen planus are best treated with triamcinolone ointment. Patients may use water-based lubricants if needed for intercourse. Topical lidocaine may also be used as-needed basis for pain relief. The natural history of most cases of lichen planus is to remit within one to two years; the exception is oral lichen planus, which often is chronic. ***(Correct Answer: c).***

This page was intentionally left blank

This page was intentionally left blank

This page was intentionally left blank

You may also try other MRCP self-assessment books, which were written by Osama S. M. Amin:

1. Get Through MRCP; BOFs:

This was published initially by the Royal Society of Medicine (RSM) Press Ltd, London in 2008. Now, the book is published by CRC Press of Taylor & and Francis. *Get Through MRCP Part 1: BOFs* provides over 600 questions and answers, allowing the reader to test their knowledge in preparation for the MRCP Part 1 examination. Questions are presented in the style used in the real examination, and answers are supplemented with useful additional explanatory material to help the reader understand why their answer was right, or wrong. The book offers a useful review of all elements of the syllabus, so the reader can feel fully prepared when they enter the examination room.

2. **Mock Papers for MRCPI, 2nd Edition**:

This was published in December 2016 by Lulu Press Inc. In this book, you will find 3 mock papers. Each one contains 100 questions in a Best of Five (BOF) format. Self-assess, try to complete each paper within 3 hours, check out your answers, read the explanation, and re-read accredited medicine textbooks to fill in the gap in your knowledge. Each question has an "objective"; try to review what the objective is about. This is the only self-assessment book specifically written to imitate the MRCPI part I examination.

3. Neurology: Self-Assessment for MRCP(UK) and MRCP(I):

This was published in September 2016 by Lulu Press Inc. You will find 792 questions of different formats, distributed into 3 chapters. Chapter 3 has many photographic materials and data interpretations themes.

4. Self-Assessment: 650 BOFs for MRCP(UK) and MRCP(I) Part I:

This was published in January 2017 by LuLu Press Inc. It provides 650 MRCP-style questions in a best of five format. A very comprehensive explanation this books supplies in order to "teach" candidates and help them get through their journey.

This page was intentionally left blank

CPSIA information can be obtained
at www.ICGtesting.com
Printed in the USA
LVOW11s2357070618
579959LV00002BB/479/P